NEOCLASSICAL HOTEL
新古典风格酒店

马竹音 编　代伟楠 译

辽宁科学技术出版社

Vernacular Classicism, a Glocal Way to Interiors

地方古典主义——室内设计的全球趋势

Vernacular architecture is a term used to categorise methods of construction, which use locally available resources and traditions to address local needs and circumstances. Vernacular architecture tends to evolve over time to reflect the environmental, cultural and historical context in which it exists. Classicism, refers generally to a high regard for classical antiquity, as setting standards for taste, which the classicists seek to emulate. The art of classicism typically seeks to be formal and restrained: of the Discobolus Sir Kenneth Clark observed, "if we object to his restraint and compression we are simply objecting to the classicism of classic art. A violent emphasis or a sudden acceleration of rhythmic movement would have destroyed those qualities of balance and completeness through which it retained until the present century and its position of authority in the restricted repertoire of visual images." The evolution of interior decoration themes has now grown to include themes not necessarily consistent with a specific period style allowing the mixing of pieces from different periods. Each element should contribute to form, function, or both and maintain a consistent standard of quality and combine to create the desired design. A designer develops a home architecture and interior design for a customer that has a style and theme that the prospective owner likes and mentally connects to. A style, or theme, is a consistent idea used throughout a room to create a feeling of completeness. Styles are not to be confused with design concepts, or the higher-level party, which involve a deeper understanding of the architectural context, the socio-cultural and the programmatic requirements of the client. These themes often follow period styles. A trend in thinking in the later parts of the 20th century, influences the ideologies of making project in general. Contextualism is centred on the belief that all knowledge is "context-sensitive". This idea was even taken further to say that knowledge cannot be understood without considering its context. The functional and formalised shapes and spaces of the modernist style are replaced by diverse aesthetics: styles collide, form is adopted for its own sake, and new ways of viewing familiar styles and space abound. Perhaps most obviously, architects rediscovered the expressive and symbolic value of architectural elements and forms that had evolved through centuries of building, which had been abandoned by the modern style. New trends became evident in the last quarter of the 20th century as some architects started to turn away from modern Functionalism, which they viewed as boring, and which some of the public considered unwelcoming and even unpleasant. These architects turned towards the past, quoting past aspects of various buildings and melding them together (even sometimes in an inharmonious manner) to create a new means of designing buildings.

"地方建筑"用于形容利用当地资源和传统方式来满足当地需求和条件的建造方式。地方建筑随着时间而进化，反映了其所处的环境、文化和历史背景。"古典主义"指具有高度古典价值的设计，为古典主义者奠定了模仿的标准。古典主义的艺术寻求条理和克制，正如肯尼斯·克拉克爵士所说："如果我们反对他的克制和浓缩，我们就是在反对古典主义艺术。律动中的重音或突然的加速将破坏古典主义延续至今的平衡而完整的特性和不断重复的视觉形象。"室内装饰主题如今已经进化到无须与特殊历史时期的风格相一致，能够融合各个不同时期的风格。各个元素可以都用在实用功能和造型装饰上，保持一致的风格，共同组成理想的设计。设计师将为委托人打造他们所喜爱的住宅建筑和室内设计。风格，或主题，是一个空间内统一的设计理念，打造出一种完整感。风格与设计理念不同，也不是更高层的建筑派别。它蕴含着对于建筑环境、客户的社会文化和项目需求的更深层次的理解。这些主题通常具有时代风格。20世纪末期的一种思想风潮影响着项目制作的整体理念。文脉主义将重心放在"所有知识都与所处的环境相关"。这一理念甚至被进一步发展为"如果没有环境，就不会理解知识"。现代主义风格中规中矩的造型和空间设计被多元化的美学所取代：风格相互碰撞，造型拥有独立的风格，到处都充满了熟悉风格和空间的新视野。而最明显的是，建筑师重新发现了建筑元素和造型的表现价值和象征价值。这些元素通过几百年的建筑一直在进化，但是却被现代风格所抛弃。新风潮在20世纪的后25年表现得尤为明显。一些建筑师开始重新回归现代实用主义。在从前，实用主义一直被设计师视为无趣，并且被一些民众认为不友好乃至讨人厌。这些建筑师回归过去，引用过去各种建筑的风格并将他们融合在一起（有时甚至不太和谐），以打造全新的建筑设计方式。

Roberto Murgia

罗伯托·马尔吉亚

CONTENTS
目录

GRAND HOTEL VILLA CORA

柯拉别墅大酒店

Completion date (项目建成时间) : 2010 (restyling) / Location (项目地点) : Florence,
Italy / Designer (设计师) : Marianna Gagliardi Architetto /
Photographer (摄影师) : Massimo Listri /
Area (室内面积) : 4,950 sqm (total square footage) + 11,000 sqm (garden)

Located in a park amid planetree woods and the Viale dei Colli gardens, few steps away from Boboli and the Oltrarno, Grand Hotel Villa Cora was built in the second half of the nineteenth century and was called Villa Oppenheim, as it belonged to this family. The project was created by the Florentine Architect Pietro Comparini Rossi, Giuseppe Poggi's pupil. The villa was built between 1870 and 1872, and the interior designer was the Engineer of Turin, Edoardo Gioja.

It consists of a monumental ground floor; the main floors are the first floor, the second and the third floor. In all the floors, except the ground floor, there are in total 46 bedrooms and suites.

On the ground floor in the Renaissance Hall there is the coffee room, the tea room, the Mirror Hall, the Moorish Hall, and the White Hall. The ground floor was mainly a matter of restoration, bringing up the ancient decorations on the wall and the restoration of the wooden floors.

On the first floor there are the historical bedrooms and suites, and all of them have ceiling frescos. All the bedrooms are provided with bathrooms decorated with "Calacatta oro" marble, enamelled steel baths and Italian taps and fittings.

In the central hall there is a very important structural reinforcement work. On the second floor all decoration are in white and gold. The pattern is fabric roses. On the third floor the old wooden staircase has been replaced with a marble one. All decoration has an Oriental style.

The restaurant Pasha is located at the basement, and the name Pasha takes its origin from "Isma'il Pascià", who stayed there in 1879. At the basement, there is a café and spa resort.

Near the villa there is a small building "Villino Eugenia", equipped with 14 bedrooms. The name takes its origin from the French empress Maria Eugenia de Montijo Bonaparte, who had lived there for nine months since 1876.

> Gioja's delicate circular, domed foyer, which is full of stuccowork, gold and frescoes, leading to the reception rooms.

> 酒雅设计的前厅为精致的圆拱形结构，采用灰泥制品、金饰和壁画作装饰，直通接待室。

> Basement plan
> 地下室平面图

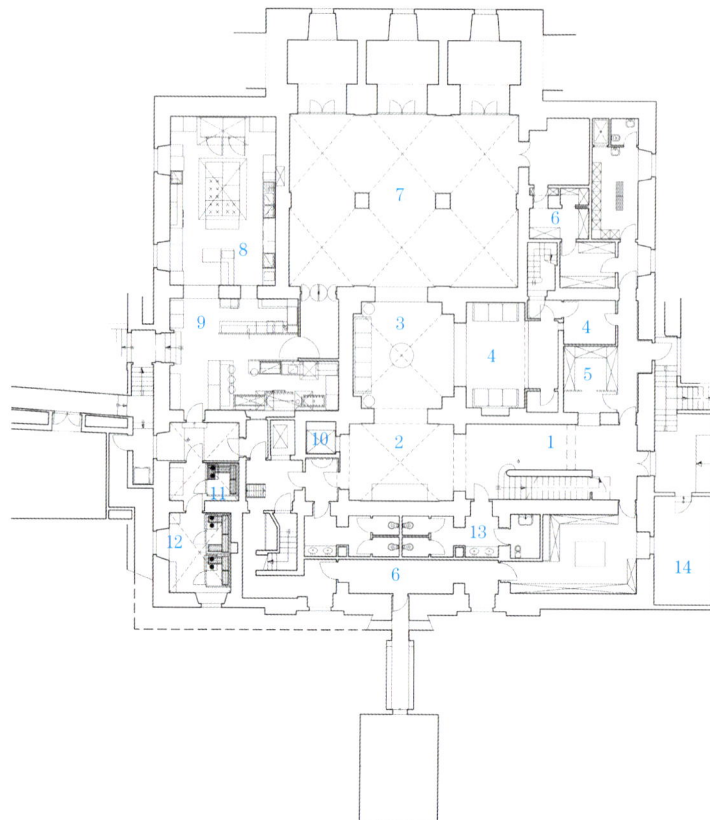

1. Main entrance	1. 主入口
2. Foyer	2. 门厅
3. Reception hall	3. 接待大厅
4. Bar	4. 酒吧
5. Cloak room	5. 行李寄放处
6. Storage	6. 仓库
7. Restaurant	7. 餐厅
8. Kitchen	8. 厨房
9. Office	9. 办公室
10. Elevator	10. 电梯
11. Refrigerator	11. 冷藏库
12. Pantry	12. 食品储藏室
13. Toilets	13. 厕所
14. Technical room	14. 技术室

> The corridor is decorated with marble floor.
> 走廊采用大理石地板作装饰。

柯拉别墅大酒店是一个公园中的别墅酒店，周围生长着郁郁葱葱的悬铃木属植物，四周被希尔斯大街的几个花园包围，距离波波利庭园和奥尔特拉诺仅几步之遥。柯拉别墅大酒店始建于19世纪后半叶，当时它归奥本海姆家族所有，因此叫做奥本海姆别墅。该酒店由弗洛伦萨建筑师Pietro Comparini Rossi（建筑师朱塞佩·波吉的学生）建造。奥本海姆别墅于1870年施工，1872年竣工，室内设计师为都灵的建筑工程师埃多奥多·酒雅。

酒店的一层是一个极为庞大的标志性空间，酒店的主要楼层是二楼、三楼、还有四楼。除了一层以外，所有楼层共分布了46间卧室与套房。

一层的文艺复兴大厅中，有一间咖啡厅、一间茶室、一个镜子大厅、一个摩尔式大厅和一个怀特大厅。一层主要采用复古的风格，墙面采用了古典装饰，并安装了古典的木质地板。

一层空间是一些历史性的客房与套房，所有客房的天花板都采用装饰画装饰。所有客房都带有浴室，浴室采用"Calacatta oro"大理石装饰，还配备搪瓷冲压钢板浴缸和意大利的水龙头与小配件。

中央大厅有一个加固的结构框架。三楼的主色调为白色和金色，并采用织物玫瑰装饰。四层老式的木质楼梯被替换成了大理石楼梯。所有装饰都体现了东方装饰风格的神韵。

帕夏餐厅位于地下室中，餐厅以伊斯梅尔·帕夏将军的名字命名，他曾于1879年下榻该酒店。地下室还有一个咖啡厅和一个水疗度假村。

柯拉别墅大酒店附近还有一个叫做"维利诺·欧仁妮"的小型建筑，建筑内有14间客房。这栋小型建筑以法国皇后玛丽娅·欧仁妮·德·蒙蒂诺·波拿巴的名字命名，她曾于1876年下榻于此，并入住长达9个月。

> The Byzantine Room inspired by the East has exquisitely carved wooden moldings.
> 受东方装饰风格的启发，拜占庭会议室采用精致的木质雕刻橡材作装饰。

> Ground floor plan
> 一层平面图

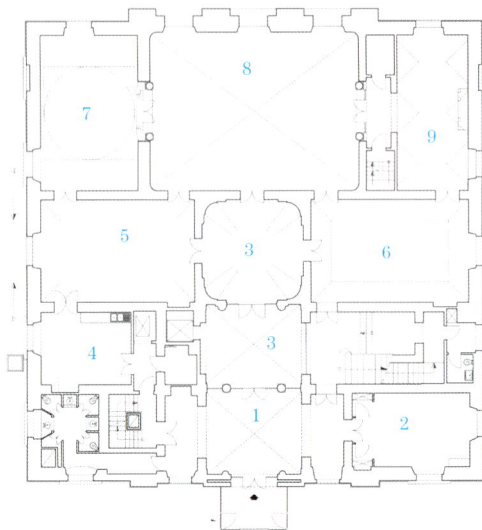

1. Entrance   1. 入口
2. Head office   2. 总部办公室
3. Foyer   3. 前厅
4. Office   4. 办公室
5. Ceramics Hall   5. 陶瓷大厅
6. Byzantine Hall   6. 拜占庭大厅
7. Moorish Hall   7. 摩尔大厅
8. Mirror Hall   8. 镜子大厅
9. White Hall   9. 白色大厅

> The Mirrors Hall is the biggest and the most majestic, versatile for any type of event, thanks to the charm of the original baroque decorations.

> 镜厅中采用古典的巴洛克艺术风格装饰，散发着迷人的气息，是最大、最辉煌的大厅，适合举行各种重大活动。

> The decorative pattern on the wall is exquisite.

> 墙面的装饰花纹十分精致。

> The wooden door is elegant.

> 木门更显优雅。

> The White Sitting Room is so-called for its exquisite
  carved white Carrara marble fireplace.

> 客厅中装饰着精致的白色卡拉拉大理石雕刻壁炉，
  因此得名白色客厅。

> Ceramique room in Le Long Bar is an intimate space.

> Long酒吧中的Ceramique厅拥有十分亲密的氛围。

> Le Long Bar has a huge colour-changing glass table.
> Long酒吧拥有一个巨大的变色玻璃桌。

> The Le Pool Bistrot restaurant is clean and tidy, and has a view of garden.
> The paintings in the restaurant were carried out by well-known artists.
> The entrance of restaurant features green velet sofa.

> Le Pool Bistrot 餐厅干净、整洁，餐厅中可一览花园的美丽景色。
> 餐厅中的壁画出自著名的画家之手。
> 餐厅入口摆放着一张绿色的丝绒沙发。

> The restaurant Pasha is full of hollow patterns.

> Pasha餐厅到处都装饰着空心图案。

> The restaurant Pasha has arched door.
> Pasha餐厅采用拱门设计。

> The Imperial Suite is dedicated to the Japanese Emperor Hirohito.

> The interior of the suite is opulent in the extreme, but by no means over-blinged.

> Deluxe rooms on the first floor are spacious.

> 皇室套房是专为日本裕仁天皇设计的套房。

> 套房室内采用极为奢华的装饰，但是一点也不夸张。

> 位于一层的豪华套房十分宽敞。

> Every room has at least one piece of antique furniture.

> 每个房间都至少采用一套仿古家具作装饰。

> The suite has warm light.

> 套房中光线柔和、暖意融融。

> Elegant curtain on the wall and queen-size bed make the suite more comfortable.

> 墙上挂着的优雅窗帘和大号卧床使套房看起来更加舒适。

> The round lamp creates warm and fragrant environment.

> 圆形的壁灯营造了温馨的氛围。

> The large wooden door is connected to the bathroom.
> 巨大的木门直通浴室。

HOTEL PALACE, BARCELONA

巴塞罗那皇宫酒店

Completion date（项目建成时间）: 2009 / Location（项目地点）: Barcelona, Spain /
Designer（设计师）: Lluis Domenech i Montaner /
Photographer（摄影师）: LHW / Area（室内面积）: 7,000 sqm

Set in the heart of the city centre, close to Las Ramblas and in the most prestigious shopping area, it is surrounded by boulevards, theatres and main banks.

It was inaugurated at the beginning of the century and has enjoyed great international prestige. The property successfully combines elegance with the most modern facilities, as well as offering first-class service.

The renovation has preserved the classic spirit that has always characterised this hotel as a distinguished symbol of Barcelona's hotel tradition. The beauty and majesty of this stunning jewel has been enhanced by a new room design, a redecoration of the public areas and lounges, and the realignment and optimisation of the architectural structure. All guestrooms and suites are equipped with last-generation technology.

This majestic building is particularly beautiful at night, with its white façade and lovely entrance. The main hall is breath-taking, with a marble floor, columns and high ceiling. There will also be a newly-opened spa providing all types of beauty treatment together with a water area, a sauna and fitness room, for future guests to benefit from the calm, serenity and luxury of unique sensations for the body and mind.

The hotel features the Caelis Restaurant, with a full offer of top-class cuisine, together with other bars and restaurant such as the Caesar Restaurant located at the hall of the hotel and the emblematic Scotch Bar.

After the restoration, the hotel will open 125 new rooms, 42 of which will be suites and junior suites, and all equipped with last-generation technology.

The Classic Deluxe rooms of 38 square metres afford the exterior city view to Gran Via Avenue. Most of them offer decorative fireplace and elegantly and charmingly classic renovated room but with a fresh look as well. The new bathroom is designed with a bathtub and a separated shower with a private closet.

> The luxurious hardwood fittings make for welcoming, cosy surroundings at the bar.
> 奢华的硬木家具打造了惬意、友好的酒吧氛围。

巴塞罗那皇宫酒店占据巴塞罗那市中心的核心地理位置，毗邻兰布拉斯大道，地处最繁华的商业区，被林荫大道、剧院和各大银行包围。

酒店于21世纪之初向公众开放，享有较高的国际声望。该酒店不仅将古典优雅与最现代的设施相结合，而且提供给游客一流的服务。

该酒店以古典装饰为特色，形成了巴塞罗那酒店传统的一个突出象征。酒店的翻新仍然保留了原古典特色。全新的房间设计、公共区和大厅的重新装修以及建筑结构的改组与优化使这座明珠酒店更加美丽、更加壮观。所有客房和套房都配备最新型的技术设备。

夜晚，这栋宏伟的大楼尤为美丽——白色的幕墙、可爱的入口。大礼堂采用大理石地板、栏杆和高高的天花板作装饰，堪称美妙绝伦。不久，水疗中心还会向游客开放，水疗中心包括温水区、桑拿房和健身室，届时将提供给游客各种各样的美容服务，游客将在宁静、祥和与奢华的环境中身心得到彻底放松。

巴塞罗那皇宫酒店的Caelis 餐厅提供各色顶级美食，还有位于酒店礼堂的Caesar餐厅和具有代表性的苏格兰酒吧等各色酒吧与餐厅。

巴塞罗那皇宫酒店经翻新后将拥有125间新客房，其中42间为套房和标准套房，所有客房都配备最先进的技术设备。

古典豪华客房占地38平方米，客房内可一览格兰维亚大街的美丽景色。大部分客房都配备装饰性的壁炉与优雅与美丽的房间，房间经翻新后既具有古典主义色彩，又呈现出全新的面貌。新浴室配备一个浴缸和一个独立的淋浴器，还有一个私人衣柜。

> The hall is decorated with golden furnitures.

> 大厅中装饰着金色家具。

> The golden mirror is luxury and elegant.

> 金色镜子，彰显奢华和优雅。

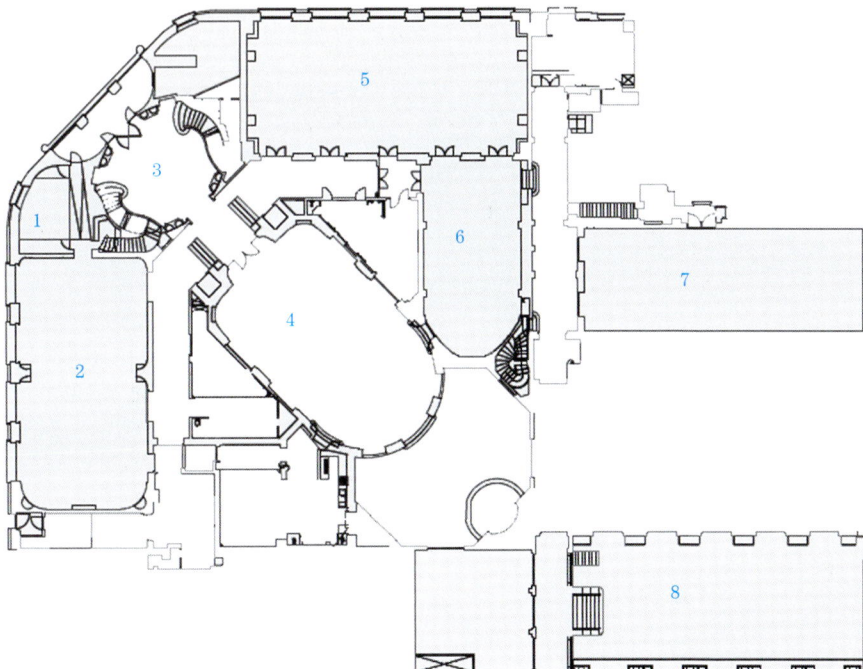

> Meeting room plan

> 会议室平面图

1. Rubí Salon
2. Caelis Salon
3. Entrance
4. Lobby
5. Gran Vía Salon
6. Dalí Salon
7. Real Salon
8. Jardín Salon

1. Rubí沙龙
2. Caelis沙龙
3. 入口
4. 大堂
5. Gran Vía沙龙
6. Dalí沙龙
7. Real沙龙
8. Jardín沙龙

> AE Restaurant is decorated in the style of a typical Parisian brasserie.

> Italian and French mirrors of the nineteenth and early twentieth century, along with silver objects belonging to the historical collection of the hotel, make a comfortable space.

> Cugat Salon features blue velvet sofa.

> AE餐厅采用经典的巴黎啤酒店的风格装饰。

> 19世纪及20世纪初的意式风格与法式风格的镜子同酒店珍藏的银器共同打造了一个舒适的空间。

> 库加特沙龙内摆放着蓝丝绒沙发，十分醒目。

> The Classic Deluxe room is exterior with abundant natural light, and completely soundproof.
> 古典豪华客房拥有充足的自然光线，并采用隔音设备，仿佛完全置身于室外。

> The Classic Deluxe is equipped with an elegant decorative fireplace, which dates from 1919.
> Spacious room, fully renovated and furnished in classic style, is very comfortable.
> The Dali Suite has taken the concept used in the halls of the hotel to achieve the refinement
  as that of The Palace with great history and prestige.

> 古典豪华客房内装饰着一个优雅的壁炉，壁炉的历史可追溯到1919年。
> 经过彻底翻新的卧室，十分宽敞，采用古典主义风格装饰，极为舒适。
> 达利套房采用酒店大厅的设计风格，打造了如古代辉煌宫殿般的优雅氛围。

HOTEL SACHER VIENNA

维也纳萨赫酒店

Completion date（项目建成时间）: 2010 / Location（项目地点）: Vienna, Austria /
Designer（设计师）: Pierre-Yves Rochon / Photographer（摄影师）: LHW /
Area（室内面积）: 8,400 sqm

Situated in the pulsative cultural heart of Vienna - directly opposite the state opera, next to the Albertina gallery and moments away from the most impressing sights of the city - the Hotel Sacher Vienna could class itself as one of them. Entwining impressive classicism of the end century and contemporary style, the Hotel Sacher Wien is the synonym for luxury and high–class standard in Vienna.

Extensive renovation work has been carried out on the Anna Sacher restaurant - now radiant in bright green pleasantly offset by black furniture with golden highlights. This interior design merely serves to underscore the splendour of the paintings by Anton Faistauer hanging in the restaurant and the priceless Lobmeyr crystal chandeliers. This interior contrasts starkly with the straight, unpretentious line taken by Sacher's chef de cuisine Werner Pichlmaier in his contemporary, light interpretation of traditional Austrian cuisine.

Its restaurant "Rote Bar" with live background piano music, winter garden and direct view to the state opera, becomes the living room of the actors of the Viennese culture scene.

The finest service, exclusive "Time to Chocolate" amenities, precious furnishings, imposing original oil-paintings, noble carpets and mulberry silk wallpapers in each of the 152 rooms make guests feel at home. The new top floors offer elegant and spacious rooms and suites with state-of-the-art bathrooms and magnificent terraces with stunning views of Vienna. Stylish materials lend each room its unique charm. At the Hotel Sacher Vienna, the guests have the opportunity to experience the genuine Viennese art of life.

The Hotel Sacher has always attached great importance to a balance of tradition and innovation. From October 2010 until mid November 2011 the hotel are therefore looking at a gentle renovation of the traditional hotel rooms. The focus is put on a sensible combination of classical interior design with innovative technology. The main emphasis is on the refurbishment of spacious and luxurious bathrooms. The French interior designer Pierre-Yves Rochon has been commissioned again to implement this scheme in the inimitable Sacher style.

> The adjacent "Schönbrunner Loge" offers enough space to relax with a cup of "Original Sacher Café".

> 宴会厅旁边的 "Schönbrunner Loge" 休息室拥有广阔的空间，游客可在此喝上一杯原创的萨赫咖啡，享受悠闲时光。

> Basement plan

> 地下室平面图

1. Marble hall	1. 大理石厅
2. Restaurant "Anna Sacher"	2. "安娜萨赫" 餐厅
3. Corridor	3. 走廊
4. Salon Mayerling	4. Salon Mayerling沙龙
5. Salon Metternich	5. Salon Metternich沙龙
6. Blaue Bar	6. Blaue酒吧
7. WC	7. 卫生间
8. Hotel lobby	8. 酒店大堂
9. Cloakroom	9. 衣帽间
10. Café Sacher	10. 萨赫咖啡厅
11. Entrance	11. 入口
12. Winter garden	12. 冬季花园
13. Lobby	13. 大厅
14. Restaurant "Rote Bar"	14. Rote bar餐厅
15. Reception	15. 前台
16. Salon Rendezvous	16. Salon rendezvous宴会厅

> The wood-panelled lobby has a grandeur glass ceiling.
> 木板装饰的大堂拥有华丽的玻璃屋顶。

维也纳萨赫酒店地处维也纳最具动感的文化中心，毗邻阿尔贝蒂娜美术馆，对面就是维也纳国家剧院，距离维也纳最著名的景点仅几步之遥。维也纳萨赫酒店本身就是一个美丽的景观。酒店将世纪经典与现代风格融为一体，是维也纳一家极为奢华的高档酒店。

酒店的安娜·萨赫餐厅经过大规模整修后，在黑色家具与金色装饰的衬托下，整体闪耀着绿色的光芒。室内设计主要用来衬托厨房中出自著名画家安东·法伊斯陶尔之手的壁画和价值连城的水晶吊灯。奢华的室内设计与萨赫餐厅主厨维尔纳·比奇尔梅尔的简单而低调的直线型工作环境形成了鲜明的对比。维尔纳·比奇尔梅尔对传统的奥地利美食进行了现代风格的诠释。

Rote Bar餐厅拥有生动的钢琴音乐背景和冬季花园，餐厅中可一览国家歌剧院的景色，因此那些演绎维也纳文化背景的演员们经常把这里当成起居室。

维也纳萨赫酒店共152间客房，每间客房内都拥有精致的服务、独特的"Time to Chocolate"设施、珍贵的家具、美轮美奂的经典油画作品、奢华的贵族地毯和桑蚕丝壁纸，给游客一种宾至如归的感觉。新增的顶层空间配备优雅而宽敞的客房与套房，并带有国际水准的浴室和壮观的阳台，阳台上可一览维也纳的美丽景色。时尚的材料将每间客房都装点得格外迷人。在维也纳萨赫酒店，游客可以体验到真正的维也纳生活艺术。

维也纳萨赫酒店一贯注重传统与创新的平衡。从2010年10月到2011年11月中旬，酒店就致力于传统客房的部分翻新工程。项目翻新的宗旨是古典的室内设计与创新技术的理性结合。项目的核心是对宽敞而奢华的浴室进行的整修工程。法国室内设计师皮埃尔·伊夫·罗雄再一次受委托，负责这个萨赫风格项目的室内设计。

> Following extensive renovation work, the Anna Sacher Restaurant is now radiant in bright green, pleasantly offset by black furniture with golden highlights.
> The Salon Maria Theresia is equipped with state-of-the-art lighting technology.
> Café Sacher Wien features its red furnitures and wall.

> 安娜萨赫餐厅经过大规模的整修后，采用点缀着金色装饰的黑色家具，在黑色家具的衬托下整个餐厅散发着明亮而愉快的绿色光芒。
> 玛丽亚·特莱西亚宴会厅采用先进的照明技术设计。
> 维也纳萨赫餐厅拥有红色的家具和墙壁。

> The name of this room hails from the magnificent Mayerling silverware on display in three showcases.

> 房间的名字来自三个陈列柜上展示的华丽的梅耶林银器。

> The hotel's elaborate art collection and antiques shape the atmosphere of the hotel and allow guests to experience the traditional Viennese lifestyle.

> 酒店内精致的艺术品和古玩打造了别致的酒店氛围，使客人可以体验到维也纳传统的生活方式。

> This elegant room impresses first and foremost with its precious marble floor and its magnificent chandeliers.

> 这间优雅的餐厅以珍贵的大理石地板和华丽的枝形吊灯为特色，给人留下深刻印象。

> Conferences and negotiations in Saloon Metternich have capacities for up to 50 people, and 26 people can take a seat at its long table.

> 梅特涅大厅中的会议室和洽谈室能容纳50人，长长的会议桌可供26人入座。

> The corridor looks clean and peaceful.
> A Junior suite combines living room and bedroom, which can be optically separated from one another.
> Have a seat in inviting seating areas and enjoy the typical Sacher flair.

> 标准客房将客厅与卧室相结合，两部分在视觉上彼此独立。
> 您可在休息区小坐一会，欣赏典型的萨赫品质。
> 走廊看上去安静、祥和。

> 标准客房将客厅与卧室相结合，两部分在视觉上彼此独立。
> 您可在休息区小坐一会，欣赏典型的萨赫品质。

> The new Executive Suites encompass the very generous room size of 50 to 60 square metres and a particularly elegant decoration.
> The hotel also offers comfortable seating areas in Junior Suite.
> In accordance with the tradition of the hotel, all the room interiors are individually fitted with luxurious furniture and art work.

> 崭新的行政套房拥有十分宽敞的房间，房间面积为50到60平方米，装饰十分优雅。
> 酒店的标准客房中还提供舒适的座椅，供客人休息。
> 按照酒店的传统，所有房间都采用奢华的家具和艺术品作装饰。

HOTEL SACHER SALZBURG

萨尔茨堡萨赫酒店

Completion date (项目建成时间) : 2003 / Location (项目地点) : Salzburg, Austria /
Designer (设计师) : Carl Freiherr von schwarz / Photographer (摄影师) : LHW /
Area (室内面积) : 7,000 sqm

The Sacher Salzburg is a grand, historic hotel situated on the banks of the Salzbach River in the city centre, with magnificent views of the old town of Salzburg.

This privately owned luxury hotel combines timeless elegance and tradition with the highest standard of service and modern conveniences. After a major refurbishment the entire hotel was upgraded to offer state-of-the-art technology.

The lobby is impressive as the prime staircase leads from the lobby up through the complete hotel. It is little in size and traditional in design leading through to a coffee/bar area, which has a big seating area with rich chairs and sofas.

The famous Café Sacher Salzburg serves the original Sacher cake in a typical Austrian coffee house ambience with an elegant touch. The restaurant Zirbelzimmer, whose wooden panelling and elaborately carved ceilings are unique in Salzburg, is a popular meeting place for gourmets. The restaurant Roter Salon is located on the footpath along the Salzach River and offers beautiful views of the old town. The guest can enjoy barbecue specialities in the casual ambience of the restaurant Salzachgrill. The piano bar serves a wide range of drinks accompanied by live music.

Many of the 113 rooms and suites were recently renovated. Even though no room resembles another, they all maintain the traditional and exclusive flair of the hotel. The categories - Standard, Superior, Deluxe and the luxurious suites differ in size, though not in technological equipments and conveniences.

All rooms and suites are furnished in a highly individual fashion and offer state-of-the-art facilities. During the renovations, the colours of the interiors were painstakingly maintained to retain the traditional atmosphere of the hotel. Each room and suite has been personally decorated by Mrs. Elisabeth Gürtler, the hotel's owner, with a great love of details. Furnishings include original paintings, precious carpets, and silk wallpapers. All rooms at Hotel Sacher Salzburg offer air conditioning, a mini bar, and bathrobes. The view of the silhouette of the old town and Hohensalzburg Fortress is truly unique.

> The "Roter Salon" is located on the River Salzach promenade with a magnificent view on the spires of the Old Town.

> "Roter Salon" 餐厅位于萨尔茨河上，餐厅中可一览萨尔茨堡老城的建筑屋顶。

> Meeting room plan

> 会议室平面图

1. Hotel lobby   1. 酒店大堂
2. Salon Karajan   2. Salon Karajan宴会厅
3. Piano bar   3. 钢琴酒吧
4. Gourmet restaurant   4. 美食餐厅
5. Roter Salon   5. Roter Salon餐厅
6. Wintergarten Rechts   6. Wintergarten Rechts宴会厅
7. Wintergarten Mitte   7. Wintergarten Mitten宴会厅
8. Wintergarten Links   8. Wintergarten Links宴会厅
9. Terrace   9. 露台
10. Sacher Café terrace   10. Sacher咖啡厅露台

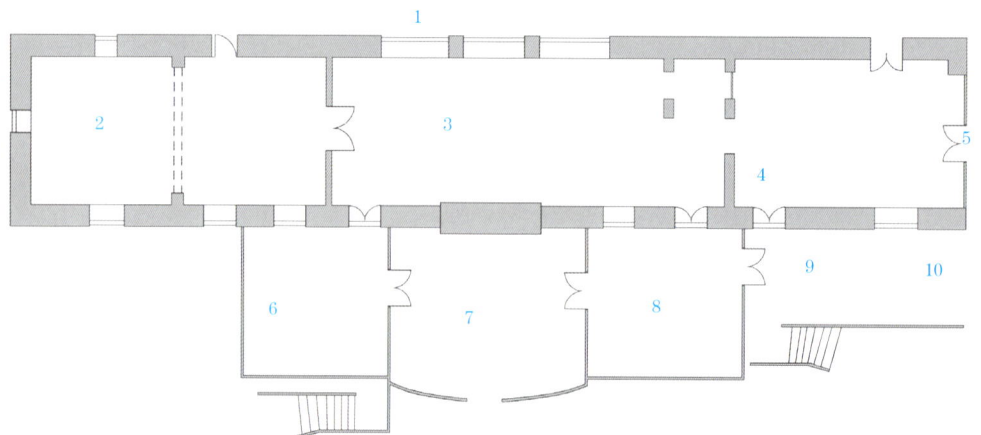

> The wooden chairs of "Roter Salon" are excellent in workmanship.
> "Roter Salon" 餐厅的椅子做工精细。

萨尔茨堡萨赫酒店是历史上一家著名的豪华酒店。该酒店位于萨尔茨堡市中心的萨尔扎河沿岸，酒店内可一览萨尔茨堡老城的美丽景色。

这家私营豪华酒店将永恒的优雅与传统同顶尖的服务和现代化的便利设施相结合。经彻底翻新后，整个酒店都提升了一个档次，提供国际先进水准的技术设施。

入口大厅内的主楼梯以大厅为起点，一直向上蔓延，贯穿了酒店的所有楼层。楼梯虽小，但却采用传统设计，沿楼梯可直接到达咖啡厅或酒吧，酒吧内摆放着奢华的座椅和沙发。

著名的萨尔茨堡萨赫咖啡厅弥漫着优雅的气息，游客可在经典的奥地利式咖啡屋内品尝到独创的萨赫蛋糕。Zirbelzimmer餐厅采用萨尔茨堡独特的木质嵌板和精心雕刻的天花板作装饰，是广受美食家欢迎的聚会场地。Roter Salon餐厅位于萨尔茨河沿岸的人行道上，餐厅内可观赏萨尔茨堡老城的美丽景色。游客可以在Salzachgrill餐厅的轻松氛围中品尝特色烧烤美食，还可在钢琴酒吧内一边欣赏音乐，一边品尝各色饮品。

该酒店共有113间客房和套房，许多客房最近都刚刚经过翻新。虽然各个客房各具特色，却也都保持了酒店独特的传统韵味。尽管各个标准客房、高级客房、豪华客房和豪华套房都配备先进的技术设备和便利设施，但各个客房的面积不同。

所有房间都采用极具个性化的装饰，并提供国际顶尖的服务设施。翻新过程中，设计师大量地保留了原酒店的色彩，旨在保存酒店的传统气息。每一间客房与套房都由酒店老板——伊莉莎白·格尔特勒亲自设计，并高度重视细节装饰。使用的装饰材料包括古典的壁画、珍贵的地毯和丝绸壁纸等。萨尔茨堡萨赫酒店的所有房间都配备空调、迷你吧和浴袍等。酒店内可观赏萨尔茨堡老城和萨尔斯堡城堡的轮廓，景色堪称无比壮观。

> Banquet room "Wintergarden" is very spacious and bright.
> The "Roter Salon" is located on the River Salzach promenade with a magnificent view on the spires of the Old Town.
> New Sacher Bar used to be a conference room before.

> 冬季花园宴会厅宽敞明亮。

> "Roter Salon"餐厅位于萨尔茨河上，餐厅中可一览萨尔茨堡老城的建筑屋顶。

> 新萨尔酒吧原来是一间会议室。

> The room is decorated with great love for detail and is adorned with valuable oil paintings and antiques.
> The living room has two large windows.
> Deluxe Room is decorated in a traditional style, with special attention to the colour.

> 客厅采用珍贵的油画和珍稀古玩作装饰，整个房间装饰十分注重细节的处理。
> 客厅拥有两扇大大的窗户。
> 豪华客房采用传统风格装饰，尤其突出了色彩设计。

> The Junior Suites are among the most popular in the hotel. They are between 40 and 50 square metres in size, and are particularly elegant.
> Superior rooms offer all the luxuries that guests would expect from a 5-star hotel.
> The suite is individually furnished with valuable oil paintings and antiques.

> 标准套房是酒店最受欢迎的客厅。标准客房面积40平方米到50平方米不等，装饰极为优雅。
> 豪华客房呈现给游客所有五星级酒店标准的奢华装饰。
> 这间套房采用珍贵的油画和珍稀古玩等个性化的装饰。

HOTEL VILLA FLORI

弗洛里别墅酒店

Completion date（项目建成时间）: 2010 / Location（项目地点）: Como, Italy
Designer（设计师）: Roberto Murgia Architetto / Stylist（造型师）: Greta Cuneo /
Photographer（摄影师）: Franco Chimenti / Area（室内面积）: 4,500 sqm

Hotel Villa Flori occupies a beautiful panoramic position right on the shores of the western branch of Lake Como, just minutes from the centre of the city. Originally an aristocratic villa, now enlarged and completely restored, the Hotel conserves the Old World charm of the 19th-century frescoes and stucco work while at the same time providing cutting-edge contemporary comforts. The project's aim was to keep in touch with the local atmosphere, local culture and local textile, and mix with the comfort of four–stars modern hotel for his international customers.

During the course of the intensive restructuring, which took two years to complete, the designer strove to maintain all the charm of the original 19th-century atmosphere, meticulously restoring the delicate stucco work and floral frescoes, the rich parquet floors and antique furnishings. But this respect for tradition did not get in the way of equally careful attention to the demands of contemporary comfort.

The physical plant is completely new, with special attentions to energy savings and low emissions. The subdivision of the spaces, while respecting the original structure, has opened up new panoramic views that turn a simple stroll through the villa into a series of continuous visual surprises. Moreover, the conversion of the attic floor made it possible to create a restful and secluded spa and fitness area.

The view is nothing short of spectacular, encompassing Torno and its distinctive bell tower, Blevio, Monte Brunate, the cupola of Como's magnificent cathedral and, further off in the distance, the tower of the 12th-century Baradello Castle.

While one might not expect to find a chapel in a luxury hotel, it has been conserved as testimony of the Villa's venerable history, and offers guests of every spiritual persuasion a beautiful setting for quiet meditation. The chapel is virtually unaltered, preserving its original marble altar and gilded wall decorations.

The Music Room was originally designated for musical training and performances, and this gracious room is now reserved for breakfast – an elegant, luminous space that invites guests to begin their day in direct contact with the natural beauty of Lake Como.

弗洛里别墅酒店坐落在科莫湖西岸，周围风景秀丽，可一览整个城市的美丽景色，而且距离市中心仅几分钟的路程。

这家酒店原本是一栋贵族别墅，现经扩建和彻底整修后既保存了以19世纪壁画与泥灰作品为特色的古典神韵，又拥有无与伦比的现代舒适。该项目旨在与当地的地域氛围和文化相呼应，采用当地的纺织品，为国际游客打造了一个舒适的四星级现代酒店。

酒店的重建工程历时两年，在高密度的重建过程中，设计师力求保存原建筑19世纪的古典特色与魅力，细心地保留了精致的灰泥作品、植物壁画和质地丰富的镶木地板，然而对古典特色的关注却丝毫没有影响对现代舒适的实现。

设计师采用了全新的电力设备，在设备选取时，着重关注了其节电与低辐射性能。空间的划分十分尊重原建筑的结构，开放的别墅与一系列连续的景观相连，打造了酒店内外通透的景色。此外，阁楼也被改造成了安静与隐蔽的水疗与健身区。

酒店周围的景色无比壮观，有Torno小镇的自然风光和镇上与众不同的灯塔，有Blevio小镇、纳特山，有科莫大教堂的圆屋顶；在远处，还有12世纪巴拉德罗城堡的塔楼。

人们可能不会期望在一家奢华酒店中发现一个小教堂，弗洛里别墅酒店的小教堂见证了弗洛里别墅庄严的历史，同时为每一个寻求精神慰藉的客人提供了一个优美的环境，他们可以在此沉思。教堂几乎没有变化，仍然保留了原来的大理石祭坛和镀金的墙饰。

音乐室起初是用作音乐训练与表演的地方，这个宽敞的房间现在是一个提供早餐的餐厅——优雅、明亮的空间邀请顾客开始他们与科莫湖自然美的亲密接触。

> Lake Como Restaurant Raimoni is decorated with subtle colours of the damask wall coverings.
> 科莫湖莱摩尼餐厅采用绸缎墙面材料的微妙色彩作装饰。

> The conference room ensures an environment of quality, taste and comfort.
> 会议室拥有高质量、高品位的舒适环境。

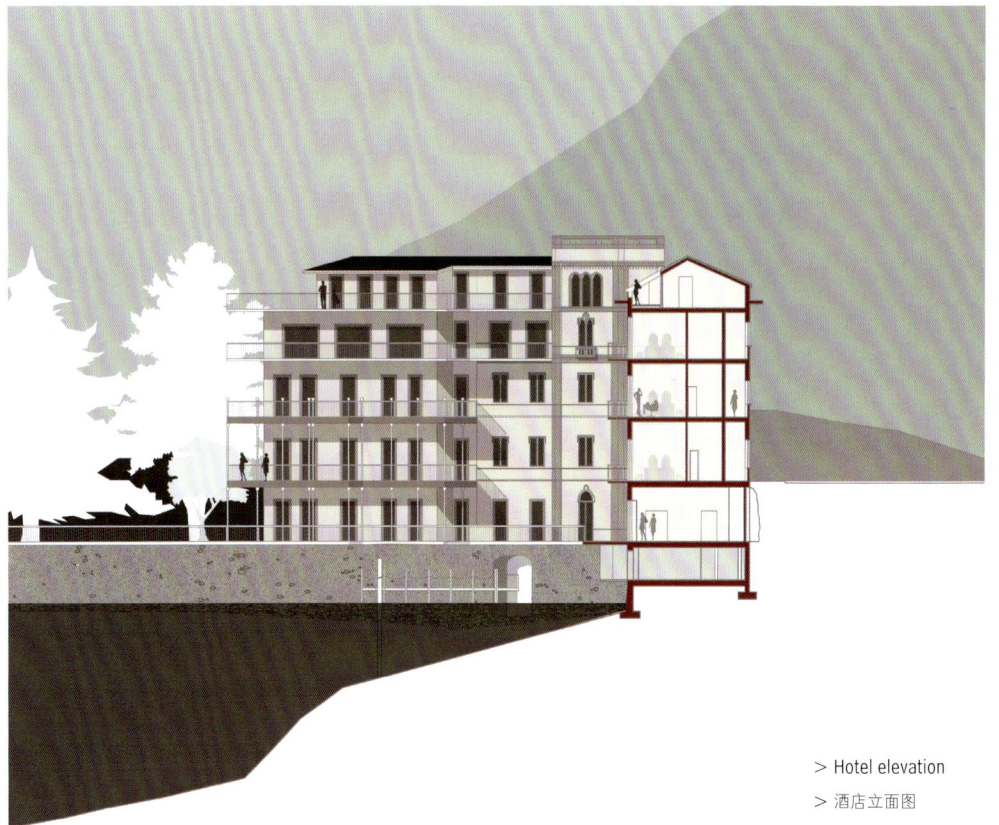

> Hotel elevation
> 酒店立面图

> A sumptuous formal dining room with recently restored the 19th-century furnishings, looks out onto the garden and lake.
> Corinthian columns make the conference room solemn and serene.
> The lobby of the hotel has a breathtaking view of Lake Como.

> 奢华的正式餐厅采用最近流行的19世纪复古式家具，餐厅中可一览美丽的湖景和花园景色。
> 科林斯石柱使会议室显得十分庄严、沉稳。
> 酒店大堂可一览科莫湖的美丽景色。

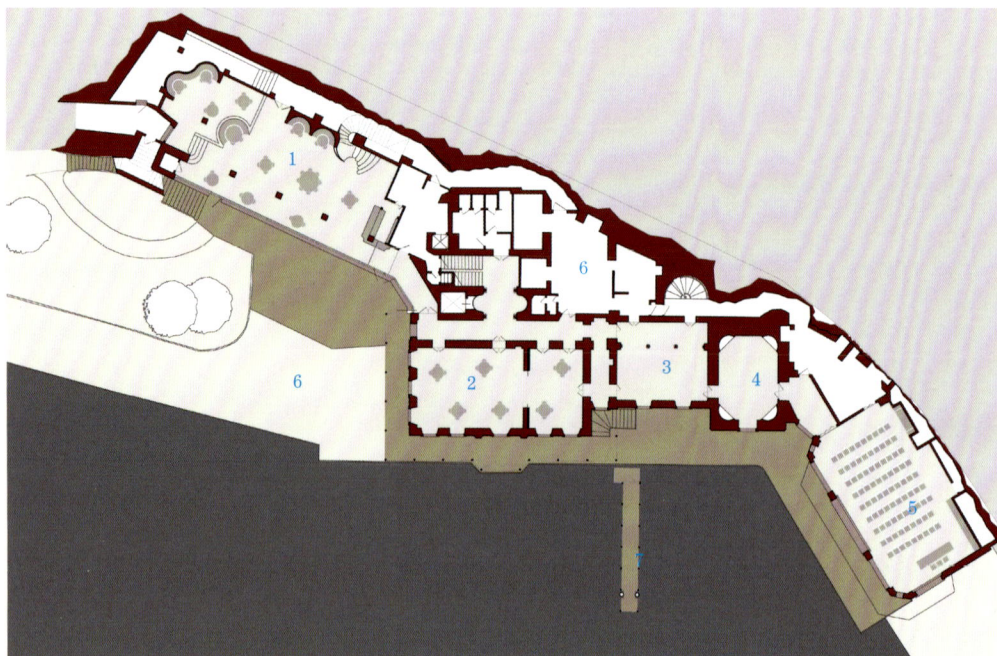

> Ground floor plan
> 一层平面图

1. Raimondi Restaurant	1. 莱蒙迪餐厅
2. Breakfast room	2. 早餐室
3. Music hall	3. 音乐厅
4. Oval room	4. 椭圆形办公室
5. Meeting room	5. 会议室
6. Garden	6. 花园

> All rooms of Hotel Villa Flori are equipped with mini bar.
> Always open on Lake Como, the wide windows have a rich variety of points of view, from terraces.
> The bedrooms are equipped with comfortable beds and graceful furnishings.

> 弗洛里别墅酒店的每一个房间都配备一个迷你吧台。
> 卧室窗户总是朝科莫湖开放，从宽敞的窗户或在露台上向外看，可饱览万千景色。
> 卧室内配备了舒适的卧床和优雅的家具。

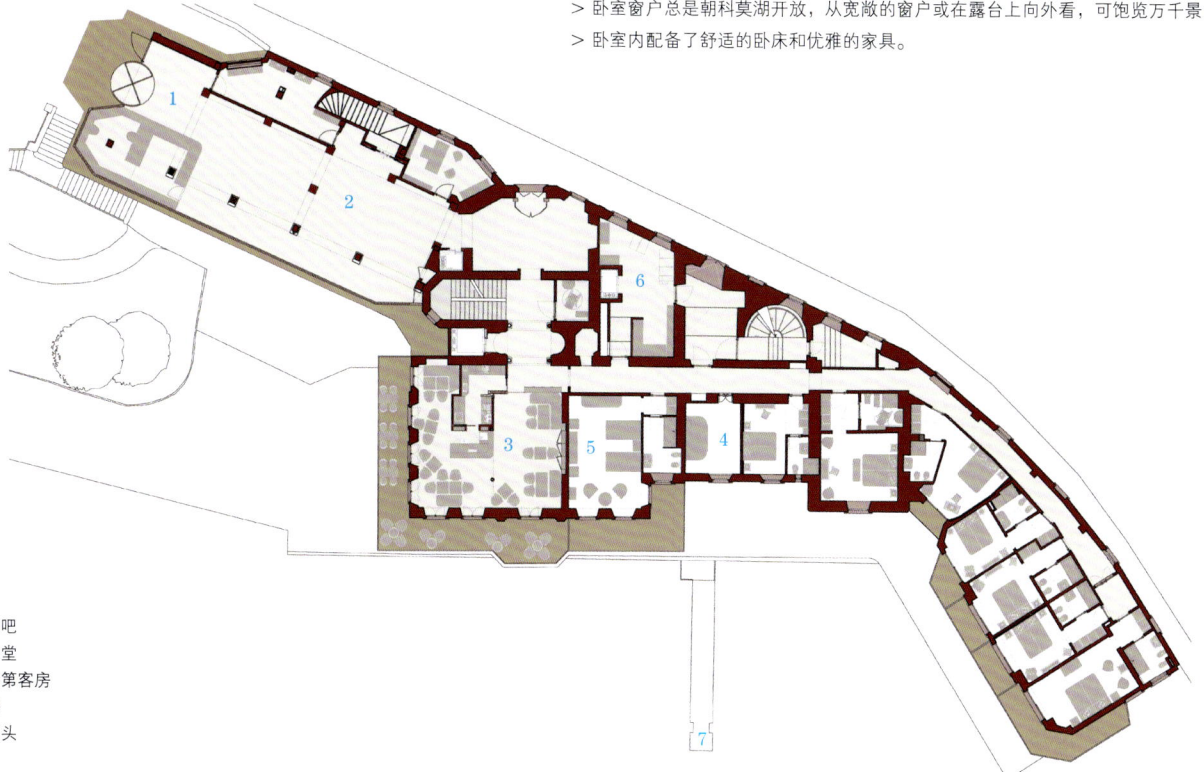

> Basement plan
> 地下室平面图

1. Reception	1. 前台
2. Hall	2. 大厅
3. Lounge bar	3. 沙发酒吧
4. Chapel	4. 小礼拜堂
5. Garibaldi Room	5. 加里波第客房
6. Laundry	6. 洗衣房
7. Boat dock	7. 小船码头

> The suite is decorated with the precious silk of upholstery and the soft colours of the floor.

> Each room has its own exclusive beauty.

> This bedroom has large windows with natural light.

> 套房采用垫衬物的珍贵丝绸和地板的柔和色调作装饰。

> 每间客房都有不一样的美。

> 宽敞的窗户保证了卧室内的自然采光。

FAENA HOTEL + UNIVERSE

费亚娜环球酒店

Completion date（项目建成时间）: 2005 / Location（项目地点）: Buenos Aires, Argentina / Designer（设计师）: Philippe Starck and Alan Faena / Photographer（摄影师）: Nikolas Koenig

Situated next to an ecological reserve in the centre of Buenos Aires and in close proximity to the historic San Telmo neighbourhood, Faena Hotel + Universe offers a wide range of outdoor activities within walking distance of many of the city's main attractions.

The interior design is the result of a close collaboration between Philippe Starck and Alan Faena and is one of Starck's most classical projects making use of a wealth of high–quality natural materials, including lapacho wood and arabasceto marble, in a style that echoes a re-birth of the Belle Epoque. Imperial furniture with gold-clawed feet, red and gold velvet curtains and fabrics and tall crystal mirrors incorporate traditional Argentine themes and patterns.

Dining options range from casual to elegant and kitchens are overseen by award-winning head chef Mariano Cid de la Paz. Gold candelabras and red Baccarat crystal rest on pure white tablecloths. White leather sofas and chairs sit atop, while plush red carpets continuing the horn theme. The moulded ceiling of El Bistro – with gold details and breathtaking chandeliers – is a recreation of the great patisseries of the 1900s.

The Library Lounge offers a relaxed atmosphere where guests may enjoy a cup of morning coffee, a healthy lunch, an informal meeting or personal time throughout the daytime and evening. Interior design items echo that of a classic estancia and include leather sofas and imperial style furniture, leather and gold–leafed bar stools, antique lamps, chandeliers and red silk curtains. Bison heads on red velvet adorn the outside wall.

100 guestrooms are spacious with a variety of rooms and suites all of which offer views of Rio de la Plata on one side or the docks and city centre on the other. Faena Hotel + Universe has ample offering of unique and spacious suites including the Presidential Suite, Park Suite, Duplex Suite, Tower Suite, Imperial Suite and the ultimate: the F Suite - the most exclusive and expensive suite in Latin America that is located on the 6th floor with sensational panoramic views of Rio de la Plata and the city of Buenos Aires.

> A pool with a great deal of personality is located in the midst of a terraced garden where guests are able to relax while lounging poolside.

> 梯台式花园的中心是一个独具特色的池塘，游客可一边在池边漫步，一边享受休闲时光。

费亚娜环球酒店坐落于布宜诺斯艾利斯市中心的一个生态保护区，毗邻历史上著名的圣太摩区，从酒店步行即可到达布宜诺斯艾利斯的主要景点。该酒店还在风景区附近提供一系列的户外活动。

酒店的室内设计是菲利浦·斯塔克与艾伦·费亚娜密切合作的结晶，也是斯塔克最经典的作品之一。斯塔克采用重蚁木和arabasceto marble等大量的高品质天然材料，再现第一次世界大战前那些美好的时代。气派的皇室家具采用金色的爪形柜脚作支撑，搭配红色及金色丝绒窗帘、织物和高高的水晶镜子，并融合了传统的阿根廷主题与图案。

酒店餐厅提供从休闲餐到正餐等一系列经典美食，

餐厅由曾获得大奖的厨师长马里亚诺·希德·德拉帕兹掌管。长毛绒地毯凸显出奢华的主题，地毯上方摆放着白色皮革沙发和座椅。El Bistro的定制天花板采用金色细节装饰，上面挂着华丽的枝形吊灯，营造出了20世纪法式蛋糕店的古典氛围。

图书馆休息室拥有轻松的氛围，游客可在此享用早茶、营养午餐，举行小型会议，或者整日地享受私人时光。室内设计仿照了古典大庄园的布置，摆放了皮革沙发、皇室风格的家具、皮革凳子、金叶状的酒吧高脚凳、古色古香的台灯、枝形吊灯和红色的丝绸窗帘等。外墙采用红色天鹅绒作装饰，上面挂着北美野牛的牛头。费亚娜环球酒店共有100个

不同类别的客房与套房，所有房间都十分宽敞，房间的一面可一览拉普拉塔河的美丽景色，另一面可观赏到码头和市中心的繁华景象。费亚娜环球酒店提供造型别致、宽敞明亮的套房，其中包括总统套房、公园套房、双层套房、城堡套房、皇室套房和顶级F套房。F套房位于酒店6层，是拉丁美洲最别致、最宽敞的套房，套房中可观赏到拉普拉塔河和布宜诺斯艾利斯的全景。

> The red chandelier echoes the red carpet.

> 红色吊灯与红色地毯相互呼应。

> The red chairs are elegant.

> 红色座椅，十分优雅。

> The stunningly white El Bistro, accented with gold and splashes of red, is an ideal destination for sophisticated, modern cuisine.
> The cellar has the largest collection of wines, which are housed within a space whose décor includes brick walls, massive pillars, and a grand chandelier.
> The sophisticated dishes are balanced by a rustic ambience of wooden floors, brick walls, and an adobe mud oven.

> 迷人的白色El Bistro餐厅，采用金色和红色的装饰做点缀，是享受绝佳现代美食的理想场所。
> 酒窖采用红砖墙、巨大的柱子和华丽的吊灯作装饰，里面收藏了各种美酒。
> 木质地板、红砖墙和泥炉灶营造了纯朴的气息，与精致的菜肴相协调。

> The Library Lounge is popular with champagne-quaffing hotel guests.

> 那些畅饮香槟酒的游客都喜欢在图书馆的休息室中享受香槟带来的乐趣。

> The Library Lounge offers a homely, relaxed atmosphere for guests to enjoy those unique moments.

> 图书馆的休息室拥有如家一般的轻松氛围，游客可在此享受别致时光。

> The refined Library Lounge is in fact decorated in exquisite style.
> The Library Lounge, seductively lit by crystal chandeliers, comfortably furnished with red velvet sofas, encourages a casual, low-key attitude.
> The walls of the spa are all decorated with brick.

> 精致的图书馆休息室装饰风格十分别致。
> 图书馆休息室中的水晶吊灯散发出性感诱人的灯光，红色的丝绒沙发无比舒适，体现了一种随意而又低调的人生态度。
> 水疗中心的墙壁全部为砖墙。

> The stuffed antelope heads are the unique artwork in Library Lounge.
> The living room of Faena Suite is very spacious with a large sofa.
> The Presidential Suite has a separated terrace.

> 塞满填充物的羚羊头是图书馆休息室中别致的艺术品。

> 费亚娜套房的客厅十分宽敞，客厅中摆放着一张大沙发。

> 总统套房拥有独立的阳台。

> The Porte Suite features its automatic red velvet curtains and venetian blinds.
> The Premium room has a great view of the river.
> The beauty centre is decorated with white furnishings.

> 土耳其宫廷套房配备自动拉伸的红色丝绒窗帘和活动百叶窗。
> 高级客房内可一览秀丽的河景。
> 美容中心全部采用白色家具作装饰。

HOTEL DE CRILLON

法国气隆酒店

Completion date (项目建成时间) : 2009 / Location (项目地点) : Paris, France
Designer (设计师) : Sybille de Margerie, sculptor César /
Photographer (摄影师) : LHW / Area (室内面积) : 8,000 sqm

Situated in the heart of the City of Light on the world-famous Place de la Concorde, Hôtel de Crillon is steps away from the luxurious boutiques of the Faubourg St.-Honoré.

The Batailles Salon in Hôtel de Crillon is very bright, decked in grey and gold, and this room provieds an outstanding view over Palace de la Concorde. It was named after the paintings that originally lined its walls, which were later replaced with todays's mirrors. It features a box-beam ceiling, gilded ornaments, Louis XV furniture and a parquet reminiscent of the Chateau of Versailles. The whole is listed as a Historical Monument.

The Counts of Crillon's former ballroom now hosts a gastronomic restaurant, acknowledged to be one of the best dining venues in the city. The place is grand and majestic, impressive in its scope as well as in the splendour of its Louis XV decoration, performed at the time by architect Ange-Jacques Gabriel: marble marquetry make of six different varieties, sumptuous crystal chandeliers, frescoes on the ceiling.

Hotel de Crillon boasts a series of suites found nowhere else in Paris: three grand Presidential Suites, each one with an outstanding view over Place de la Concorde. Whether through its dimensions or decorations, each majestic lounge will be impressive you: a spectacular high ceiling, marble floorboards, precious silk drapings, gold-leaf ornamentation…Here is superb instance of French classicism combined with full modern comfort.

All suites are exquisitely furnished and decorated with Aubusson carpets, Baccarat chandeliers and Wedgewood medaillons. Both the Bernstein Suite and the Louis XV Suites are office equipped. The Duke de Crillon suite is decorated in lovely warm tones, using the superbly restored golden oak panels of the floors, walls and ceilings in the living room and bedroom as a backdrop for the many portraits of the hotel's most esteemed guests. Roomy and luxurious, Deluxe Suites proved high ceilings and superb "Grand Siecle" furniture, as well as a sculptural fireplace. They include a bedroom featuring a very large bed, a separate lounge, two entrances and a bathroom that is made of Carrare marble.

> Adorned with a curtain of ivy, this peaceful haven nestles at the core of the hotel's sumptuous architecture.
> Luxury lobby is decorated with checkered black and white polished floor.
> The traditional architecture continues inside with a beautiful lobby and reception.

> 这个静谧的人间天堂位于奢华的酒店建筑的核心，采用常春藤幕帘作装饰。

> 奢华的酒店大堂采用黑白格子图案的抛光地板作装饰。

> 传统的酒店建筑内部是装修精美的大堂与前台。

法国气隆酒店坐落于光明之城——巴黎市中心举世闻名的协和广场，距离奢华的香榭丽舍大街仅几步之遥。

酒店内的Batailles沙龙宽敞明亮，以灰色和金色为主色调，可观赏香榭丽舍大街的美丽景色。Batailles沙龙以原建筑墙壁的挂画命名，挂画现由镜子取代。室内主要有木梁天花板、镀金饰品、路易十五家具和法国凡尔赛宫式的镶木地板。整体外形堪称是一座历史丰碑。

气隆酒店前宴会厅现被改造成了一个美食餐厅，该餐厅是巴黎最好的进餐场所之一。餐厅装修大气辉煌，采用路易十五年代由建筑师雅克·加布里设计的饰品作装饰，无论是规模还是气势都无比壮观，其中有6种不同种类的大理石镶嵌细工装饰品、华丽的水晶吊灯和屋顶壁画。

气隆酒店还有巴黎绝无仅有的一系列套房，其中包括3间壮观的总统套房，每一间套房内都可一览协和广场的辉煌景色。套房内配备高高的天花板，天花板大气恢弘，还有大理石地板、珍贵的丝绸窗帘和金叶饰品，无论是维度还是饰品，都会让客人流连忘返，是法式经典与现代舒适相结合的完美杰作。

所有套房都装修精美，采用欧比松地毯、Baccarat水晶吊灯和威其伍奖章做装饰。伯恩斯坦套房和路易十五套房都配备办公室。气隆公爵套房以可爱的暖色调作装饰，客厅和卧室的地板、墙壁和天花板采用壮观的、有益身心的金橡木作为原材料，木质背景的套房中摆放着一尊尊画像，上面画着酒店最尊贵的客人。宽敞奢华的豪华套房内配备高高的天花板、华丽的"Grand Siecle"家具和雕刻精美的壁炉。每间套房内都有一间卧室，卧室内摆放着一张大床、一个独立的休息室、两个入口和一间以Carrare大理石为原材料的浴室。

> Very bright, decked in grey and gold, Batailles Room provides an outstanding view over Place de la Concorde.
> This superb lengthwise meeting room provides an ideal setting for conferences or major receptions.
> Les Ambassadeurs Restaurant is impressive in its scope as well as in the splendour of its Louis XV decoration.

> Batailles宴会厅采用灰色和金色的色调装饰，十分耀眼，宴会厅中可一览巴黎协和广场的美丽景色。
> 狭长的会议室拥有举行大型会议或大型招待会的理想环境。
> 大使餐厅无论在规模上还是它路易十五风格的壮观装饰上都令人叹为观止。

> Contemporary and relaxed, l'Obé Restaurant has established itself as a leading venue for stylish.
> Next to reception, there is a small gallery for guests.
> Les Ambassadeurs Restaurant is grand and majestic, with a huge terrace.

> 采用现代风格装饰的l'Obé餐厅拥有轻松的氛围，现已成为一家著名的时尚餐厅。
> 前台附近有一间专门为客人准备的小型画廊。
> 大使餐厅拥有一个广阔的阳台，十分大气、辉煌。

> 采用现代风格装饰的l'Obé餐厅拥有轻松的氛围，现已成为一家著名的时尚餐厅。
> 前台附近有一间专门为客人准备的小型画廊。

> With its "Grand Siecle" furniture, the Historical Suite beautifully represents classical Parisian style.

> The living room of presidential suite has period floorboards, precious silk drapings.

> The Deluxe Suite possesses its own charm and treasures, impressive dimensions and spaces.

> 这间历史套房采用"Grand Siecle"家具作装饰，装修十分精美，再现了经典的巴黎风格。

> 总统套房的客厅采用古老的地板和珍贵的绸缎作装饰。

> 豪华客房拥有其个性化的迷人魅力和珍贵装饰，并拥有超大的维度和广阔的空间。

> The wall of this presidential suite is decorated with wood and a painting.

> This living room of Historical Suite is full of red furniture.

> Wall, curtain, carpet and sofa are all decorated with flower pattern.

> The living room of presidential suite has a piano.

> Duc de Crillon Suite includes a separate living room, featuring fully modern comfort.

> 总统套房的墙壁采用木头作装饰，墙壁上还挂着一幅壁画。

> 历史套房的客厅全部采用红色家具装饰。

> 墙壁、窗帘和沙发都饰有花朵图案。

> 总统套房的客厅中摆放着一架钢琴。

> Duc de Crillon套房内有一间独立的客厅，充分体现了现代装饰的舒适。

> The soft lighting, the deep sofas and the comfortable wing chairs welcome guests to settle down.
> 柔和的灯光、长长的沙发和舒适的翼状背靠椅邀请客人下榻。

> Deluxe Suite provides high ceilings and superb "Grand Siecle" furniture.
> This bedroom has a french window with the view of garden.
> The room is a subtle blend of furniture in Louis XV style and decoration that is both sober and warm.

> 豪华套房拥有高高的屋顶和华丽的"Grand Siecle"家具。
> 这间卧室拥有巨大的落地窗，从窗户可一览花园的美丽景色。
> 这间客房融合了路易十五风格的家具和庄严而又温暖的装饰。

THE VILLA BY BARTON G.

巴顿G.别墅酒店

Completion date (项目建成时间) : 2010 / Location (项目地点) : Miami Beach,
United states / Designer (设计师) : Massimo Comoli / Photographer (摄影师) :
LHW / Area (室内面积) : 3,200 sqm

Located in Miami Beach, Florida, this former mansion of fashion designer Gianni Versace is an Italian-style palace with a mix of Persian, Egyptian and Baroque interior decor thrown in. Minimalists should be warned that the villa is a monument to excess and makes no apology for its opulence.

Renowned events impresario/ restaurateur Barton G. Weiss takes up style where fashion designer Gianni Versace left off after a more than $2.5 million renovation of the 1,759-square-metre property completed in March 2010. The gated Villa is Miami Beach's exclusive backdrop for weddings and events, and gives hotel guests exclusive access to the mansion's upper floors and two private lounges.

Accommodations range from the stunning Medallion Suite (40 square metres) overlooking one of the most beautiful mosaics in the mansion to the amorous Venus Suite (132 square metres) with two private balconies perched above Ocean Drive and the mansion's Thousand Mosaic Pool. The Villa Suite (109 square metres), once Versace's bedroom, boasts a nine-foot double king-size bed, two balconies, seven closets, and a custom oversized shower. Guests are welcomed by their dedicated British-trained and certified butlers who personally attend to them throughout their stay. In-room, poolside, and rooftop spa services are available, as is open access to the mansion's one private lounge, lush Mosaic Gardens, and the Thousand Mosaic swimming pool. Guests can luxuriate in one of the Villa's 10 custom suites featuring king–sized beds or custom double king beds, custom Italian marble oversized bathrooms with two shower heads, expansive closet space, separate living rooms or sitting areas and balconies or patios or enjoy exclusive access to the Thousand Mosaic Pool and two private lounges. They can also worship the sun or gaze at the stars from the Rooftop Lounge, and dine in the Villa's restaurant, set in the pebble-mosaiced Dining Room with a focus on refined continental cuisine, or choose to breakfast or lunch in suite or poolside. The Dining Room at The Villa By Barton G., with its pebble mosaiced walls and storied history, features refined continental cuisine fused with Barton G. style.

> The entrance greets guests with a pool and a terrace tiled in thousands of colourful mosaics, fountains, and frescoes.
> The stained glass in Roman palazzo style stands by the side of the pool.
> A three-storey structure with a Spanish-style inner courtyard contains the hotel's apartments.

> 酒店入口处设有一个池塘和露台，露台表面采用上万个色彩斑斓的马赛克拼接而成，并采用多个喷泉和多幅壁画作装饰。
> 池塘的一边是罗马宫殿风格的彩色玻璃。
> 酒店公寓位于一栋带内庭的三层建筑中，庭院采用西班牙风格设计。

巴顿G.别墅酒店坐落于美国佛罗里达州的迈阿密海滩上，酒店前身是由时装设计师詹尼·范思哲设计的一栋意大利风格的宫殿，该宫殿融合了波斯、埃及和巴洛克等各色风格的室内设计。对于最简派艺术家而言，酒店装饰似乎过分奢华，然而，这种奢华却又恰到好处。

继时装设计师詹尼·范思哲于2010年3月花250多万美元对1,759平方英尺的酒店进行翻新设计后，著名的艺术活动总监兼饭店老板——Barton G．Weiss继续负责酒店的风格设计。迈阿密海滩封闭的别墅是举行婚礼和重大活动的理想场地，游客可从别墅直接到达酒店的顶层和两间私人会所。

酒店客房包括美妙绝伦的银章套房（面积为40平方米），银章套房内可观赏到整栋大楼最美丽的镶嵌图案；还有风情万种的维纳斯套房（面积为132平方米），维纳斯套房带有一个私人阳台，位于海洋大道和数千块马赛克拼合成的游泳池上方。别墅套房（面积为109平方米）曾经是范思哲的卧室，拥有1张9英尺的特大号双人床、2个阳台、7个衣柜和1张特制的超大号浴室。一进入酒店，迎接客人的就是曾在英国接受培训并被授予证书的男管家，他们将为客人提供住宿全程的私人服务。酒店还提供室内、泳池边及屋顶等不同区域的水疗服务，水疗区可直通酒店的两个私人会馆、翠绿的马赛克花园和数千块马赛克拼合成的游泳池。别墅中共有10间不同特色的套房，套房配备特大号床或特制的大号双人床、意大利大理石镶嵌的超大号浴室，浴室内有两个淋浴头，还有宽敞的衣柜、独立的客厅或

起居室、阳台或露台，游客可在此尽情享受，还可以通过唯一的一条小路进入马赛克泳池和两个私人会所，也可以在屋顶休息室沐浴阳光或欣赏漫天繁星，亦可在别墅餐厅中享受美食——餐厅采用卵石拼接而成，主要以经典的欧式西餐为特色，此外，游客还可以选择在套房中或在泳池边进餐。巴顿G.别墅酒店内的餐厅墙壁采用卵石拼接而成并用历史画作装饰，主要提供欧式西餐与巴顿G.特色风格的美食。

> Softly-lit walls in the intimate restaurant are a mosaic of river rocks interspersed with stained-glass windows under frescoed ceilings.

> 诱人的餐厅中，柔和的光线照亮礁石拼接而成的马赛克墙壁，墙壁上的 窗户为不锈钢材料制成，上方是壁画装饰的天花板。

> Moroccan-style Observatory lounge overlooks the Atlantic Ocean.
> 摩洛哥风格的瞭望台俯视大西洋。

> Moroccan room plan
> 摩洛哥客房平面图

1. Entrance
2. Ballroom
3. Showers
4. Bathroom

1. 入口
2. 宴会厅
3. 淋浴设备
4. 浴室

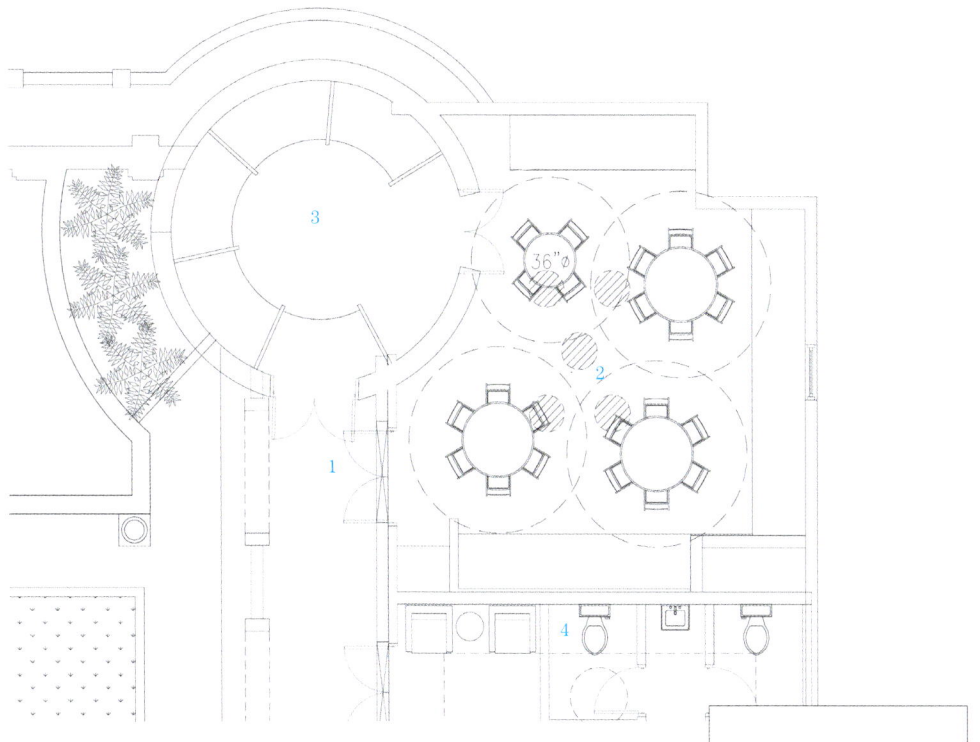

> The living room of Signature Suite has a zebra-striped sofa.

> The cosy Moroccan lounge is full of warm colour fabric.

> The Medallion Suite has Gianni Versace's signature all over it.

> 招牌套房的客厅中摆放着一款斑马条纹的沙发。

> 惬意的摩洛哥大厅中布满了暖色调的织物。

> 银章套房中到处可见詹尼·范思哲的签名。

> Mosaic Suite is decorated with a red bed and golden fabric.
> The pretty blue-and-white Azure Suite is less grand but still heart-stirringly sweet.

> 马赛克套房中装饰着一张红色的卧床和金色的织物。
> 蓝白色相间的蔚蓝套房虽不十分壮观，但却仍然甜美、诱人。

> Versace's one-time bedroom is now The Villa Suite, contains a nine-foot double king-size bed, two balconies.

> 范思哲曾经住过的卧室现命名为"别墅套房"，该套房配备一张9英尺宽的特大号双人床和两个阳台。

> Hummingbirds and doves decorate the walls of The Aviary Suite.
> 蜂鸟和白鸽装饰着鸟舍套房的墙壁。

GRAND HOTEL WIEN

维也纳大酒店

Completion date (项目建成时间) : 2005 / Location (项目地点) : Vienna, Austria
Designer (设计师) : Sign Design / Photographer (摄影师) : LHW
Area (室内面积) : 13,000 sqm

The Grand Hotel Wien is located in the heart of Vienna, on the famous boulevard "Wiener Ringstraße", and just a few steps to the Vienna State Opera and the "Kaerntner Street", which is incomparable.

With a spectacular view over the city of Vienna, the award-winning restaurant Le Ciel, located at the 7th floor of the Grand Hotel Wien, is known as one of Vienna's best gourmet restaurants. Le Ciel Restaurant is styled in warm traditional colours; the china and silver were all especially created for this legendary restaurant.

The Grand Café Restaurant is decorated in warm and friendly colours and offers 92 seats, partly with a view on the famous Vienna "Ringstrasse".

The 205 rooms and suites are spacious and designed in typical Viennese style. They guarantee luxury, comfort and convenience. Exquisite fabrics, rare antiques, classic decoration and state-of-the-art technical equipments will make guests feel thoroughly at home.

With a total of 220 square metres and a magnificent view on the famous Ringstrasse, the Presidential Suite is truly distinctive. Many international VIPs from politics, economics, art and the entertainment industry have already stayed here. The largest suite of the Grand Hotel Wien features a big entry hall, a spacious living room, a huge bedroom with a canopy bed, an elegant dining room for up to 8 guests, a study and a luxurious marbled bathroom with an integrated whirlpool in the bathtub and a separate shower.

The Grand Hotel Wien offers three Deluxe Suites, all with a total of 130 square metres and a sweeping view on the famous Ringstrasse. Furnished mainly in soft beige tones, all 3 suites welcome the guests in an entry hall with separate wardrobe, a decorative living room with open fireplace, balcony and dining table.Measuring 116 square metres, all four Senior Suites are very spacious and fulfill even the most discerning guest´s wishes. They are decorated predominantly in a soft apricot hue, which adds a very elegant and warm atmosphcre. All Senior Suites are located at corners, allowing a beautiful view on the Viennese City and letting a lot of daylight in.

> Le Ciel restaurant is styled in warm traditional colours; the china and silver were all especially created for this legendary restaurant.
> Le Ciel餐厅采用温馨的传统色调装饰，设计师还为这间传统的餐厅特意布置了一下瓷器和银器。

维也纳大酒店地处维也纳市中心著名的"Wiener Ringstraße"大街上，距离维也纳国家歌剧院及美妙绝伦的"Kaerntner大街"仅几步之遥。

位于酒店7层的Le Ciel餐厅被誉为是维也纳最棒的美食馆，Le Ciel餐厅内可一览维也纳整个城市的壮观景色。Le Ciel餐厅采用温暖的传统色调装饰，瓷器与银器的搭配更增添了这家餐厅的传奇色彩。

Grand Café餐厅采用温暖的色调装饰，给宾客一种宾至如归的感觉。Grand Café餐厅提供92个座位，餐厅内的部分区域可俯视著名的维也纳环城大道的美丽景色。

酒店的205间客房与套房宽敞明亮，以典型的维也纳风格装饰。这些客房与套房不仅装饰奢华，而且温馨舒适、服务便捷。高雅的织物、珍贵的古玩、古

典的装饰和先进的技术设备彻彻底底地为顾客营造了一种家的感觉。

与众不同的总统套房面积达220平方米，总统套房内可一览著名的环城大道的美丽景色。许多政界、经济界、艺术与娱乐界的国际贵宾曾下榻于此。维也纳大酒店内最大的客房拥有一个广阔的入口大厅、一间宽敞的客厅、一间奢华的大理石浴室，浴室内还配备一个漩涡式浴缸和一个独立的淋浴设备。

维也纳大酒店拥有3间豪华套房，各个套房的总面积达130平方米，可俯视著名的环城大道。3间豪华套房主要采用柔和的浅褐色色调装饰，迎接宾客的入口大厅带有一个独立的衣橱、一个装饰典雅的客厅，客厅内有一个壁炉，还有一个阳台和一张餐桌。

酒店的4间高级套房共116平方米，十分宽敞，可满足任何顾客的要求。4间高级套房主要采用柔和的杏色作装饰，从而为房间增添了一丝优雅与温暖的气息。所有高级套房都位于建筑的四角，既使游客能观赏到威尼斯城的美丽景色也使室内拥有良好的采光。

> There is a large sweeping staircase in the centre of the lobby.
> 酒店大堂的中心有一个巨大的弧形楼梯。

> This elegant dining room can accomodate up to 6 guests.
> 优雅的餐厅能容纳6名客人同时用餐。

> Fourth Banquet floor plan
> 五层宴会厅平面图

1. Ladder
2. Lift
3. Bathroom
4. Salon

1. 楼梯
2. 电梯
3. 浴室
4. 沙龙

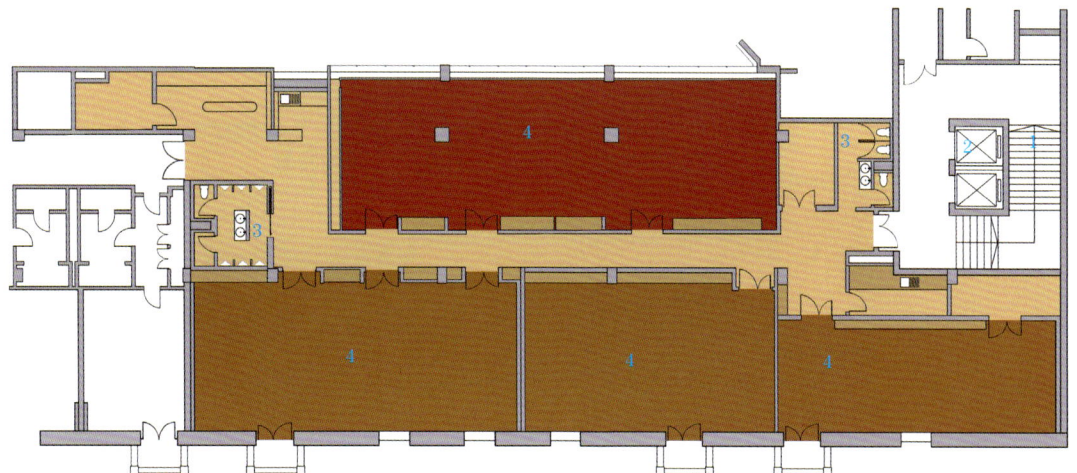

> Furnished mainly in soft beige tones, Deluxe Suite welcomes the guests in an entry hall with a decorative living room and a dinning table.
> The Royal Suite features red furnitures and red curtain.
> The decoration of Junior Suite is held in soft green colours, which gives the suite a very relaxing atmosphere.
> The Senior Suite is decorated predominantly in a soft apricot hue, which adds a very elegant and warm atmosphere.
> The Executive Suite offers their guests a lovely living area separated from the bedroom and a dining table.

> 豪华客房采用柔和的杏色作装饰，入口大厅还带有一间装饰性的客厅和一张餐桌，向客人发出盛情邀请。
> 皇室套房以红色家具和红色窗帘为特色。
> 标准套房采用柔和的绿色为主色调，营造了十分轻松的室内氛围。
> 高级套房采用柔和的杏色为主色调，杏色为房间增添了一丝优雅和温馨的气息。
> 行政套房为客人提供了一间可爱的客厅，客厅与卧室和餐厅分离。

> The Royal Suite is truly distinctive. Many international VIPs from politics, economics, art and the entertainment industry have ever stayed here.

> 皇室套房十分别致。许多政界、经济界、艺术界和娱乐界的国际贵宾都曾下榻于此。

> The bedroom of Deluxe Suite features either a large king-size bed.

> 每间豪华客房的卧室都有一张特大号的卧床。

> The residential-styled Deluxe Rooms, elegantly furnished in the colours of green and yellow, offer courtyard views.
> Each Superior Room is equipped with antique furniture, beautiful fabrics, two telephones and a flatscreen TV.
> The furniture is antique and tastefully combined with beautiful fabrics in green and yellow colours. The wallpaper is made of pure silk.

> 住宅风格的豪华客房采用绿色和黄色为主色调，精致优雅，套房中可一览庭院景色。
> 每间高级客房都配备古典家具、精美的织物、两步电话和一台纯平电视机。
> 古典家具与黄色和绿色的精美织物相结合。壁纸采用纯正的丝绸制成。

GRAND HOTEL EUROPE

欧洲大酒店

Completion date（项目建成时间）: 2009 / Location（项目地点）: St. Petersburg, Russia / Designer（设计师）: Michel Jouannet / Photographer（摄影师）: Grand Hotel Europe / Area（室内面积）: 15,000 sqm

Grand Hotel Europe is situated in the heart of St. Petersburg within walking distance of the Winter Palace Square and Hermitage Museum.

Located on the hotel's Historic Floor, each suite has its own Russian historic name, with an interior to match, and reflects the rich history of both the hotel and St. Petersburg. Themes include Pavarotti, Stravinsky, Faberge and Romanov.

All of the suites are spacious, with an area of 55 to 97 square metres, and 4.3 metre-high ceilings. Each has a vestibule, a living room, a bedroom and a large bathroom. Their windows look out onto the most picturesque spot in the historic centre of St. Petersburg – Arts Square, with its monument dedicated to the great poet, Alexander Pushkin, and the building of the Noble Assembly.

The Pavarotti Suite is the room in which the celebrated Italian tenor stayed during his final tour in 2004. This suite has always been a favourite with musicians due to the antique grand piano that stands in the living room. The interior is stylised in the spirit of the finest opera houses in the world – the Opera Garnier in Paris and La Scala in Italy. The colour scheme is dominated by hues of gold and red, and the bathroom will be finished in contrasting types of black and pink marble.

The Dostoevsky Suite is named after Fyodor Mikhailovich Dostoevsky, who was a frequent guest of the hotel. To capture the mood of this great Russian writer, the designer has chosen tones that are fresh, yet deep and serious. The walls are decorated with wallpapers featuring a 19th–century style pattern, and the living room contains a large desk for literary work.

The Imperial Yacht Suite is named after the Russian royal yacht, the Derzhava, which combines the stunned contemporaries with the unprecedented opulence of its interiors. Shades of marine colours dominate the colour scheme of the suite, while the bathroom is decorated in green and cream marble.

> The Suite consists of a large living room with a winter garden on a small, glass-covered veranda.

> 套房带有一间宽敞的客厅和一个玻璃覆盖的小型阳台上，阳台上是一个冬季花园。

欧洲大酒店地处圣彼得堡市中心，距离冬宫广场和艾尔米塔什博物馆仅几步之遥。

欧洲大酒店的所有套房都位于同一楼层，这个楼层具有深厚的历史意义。每间套房都以俄罗斯历史上著名的人物命名，并且拥有与名字相匹配的室内布置，反映了酒店和圣彼得堡悠久而灿烂的历史文化。各个套房的主题包括帕瓦罗蒂、斯特拉文斯基、法贝热和罗曼诺夫等。

各个套房面积从55平方米到97平方米不等，天花板高度都为4.3米，十分宽敞。每间套房都配备一个门廊、一个客厅、一个卧室和一个大浴室。从套房的窗户可观赏到圣彼得堡历史中心文化广场如画般的景色，还可以看到为伟大诗人亚历山大·普希金修建的纪念碑和贵族议院等建筑。

帕瓦罗蒂套房是为了纪念意大利著名男高音歌唱家——帕瓦罗蒂2004年的圣彼得堡之行而打造的客房。套房的客厅内摆放着一架古老的大钢琴，因此这里经常是音乐家们的最爱。套房的室内设计风格与世界上最精致的歌剧院风格相似，比如巴黎的加尼叶歌剧院和意大利的斯卡拉歌剧院等。色彩设计主要以金色和红色为主色调，而浴室采用黑色和红色大理石装饰，形成鲜明的对比。

陀思妥耶夫斯基套房以酒店的常客——陀思妥耶夫斯基的名字命名。为了捕捉到这位伟大的俄罗斯作家的心境，设计师选择了新鲜而又深沉严肃的色调。墙壁采用布满19世纪花纹的壁纸装饰，客厅内摆放着一张用于文学写作的大办公桌。

皇家游艇套房以俄罗斯皇家游艇the Derzhava命名，其室内设计流露出令人惊叹的现代气息和无与伦比的奢华感。一抹抹海洋色彩成为套房的主调，而浴室则以绿色及奶油色的大理石铺设。

> This suite is named after Dostoevsky to capture the mood of this great Russian writer, and the designer has chosen tones that are fresh, yet deep and serious.

> 该套房以俄罗斯大作家——陀思妥耶夫斯基的名字命名，为了衬托这位大作家的心境，设计师还特意选择了清新而又深沉、庄重的色调。

> Ground level and mezzanine level plan
> 一层及夹层平面图

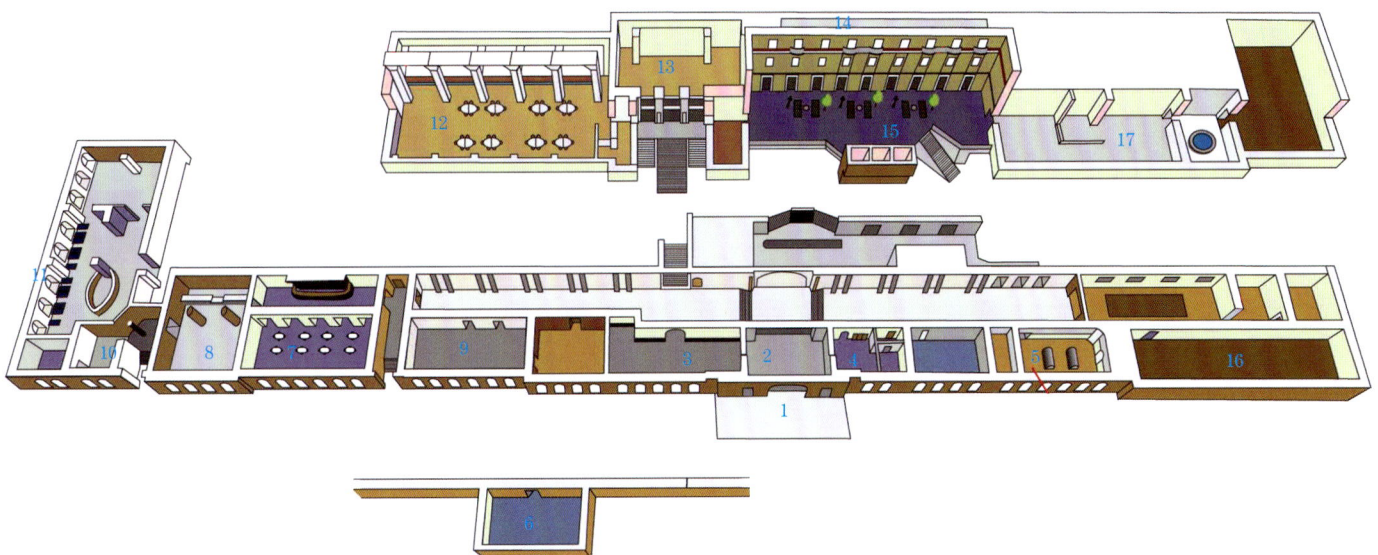

1. Entrance	10. Flower shop	1. 入口	10. 花店
2. Reception	11. Babochka Boutique	2. 前台	11. Babochka精品店
3. The lobby bar	12. C Europe Restaurant	3. 大堂酒吧	12. C Europe 餐厅
4. Macbiavelli Boutique	13. Caviar Bar & Restaurant	4. Macbiavelli 精品店	13. Caviar 酒吧与餐厅
5. Zilli Boutique	14. Art and souvenir shops	5. Zilli精品店	14. 艺术品与纪念品店
6. Smolensk Diamonds Jewellery	15. Mezzanine Cafe	6. Smolensk钻石首饰店	15. 中央咖啡厅
7. Rossi's	16. Conference facilities	7. Rossi's餐厅	16. 会议设施
8. Chopsticks	17. Mkbailov Gallery	8. Chopsticks餐厅	17. Mkbailov画廊
9. Mercury Watches & Jewellery		9. Mercury手表与珠宝店	

> The Pavarotti Suite is the room in which the celebrated Italian tenor stayed during his final tours in 2004, with an antique grand piano standing in the living room.
> The historic suite has retained all of their antique elements of décor thanks to the painstaking restoration work carried out in the hotel.
> The rooms are furnished with antique furniture from the hotel's collection.
> The historic interior and spirit of the age of Lidval are preserved in the room.

> 帕瓦罗蒂套房的客厅中摆放着一架古老的大钢琴，著名的意大利男高音歌唱家帕瓦罗蒂于2004年下榻该酒店。
> 酒店经过彻底的整修后，这间具有历史意义的套房保存了所有古典的装饰元素。
> 客房中装饰着酒店收藏的古董家具。
> 这间客房保存了 Lidval时代的精神与古典装饰。

> The Stravinsky Suite is named after the composer Igor Fyodorovich Stravinsky, whose music is associated with spring, and the interior is dominated by joyful hues of spring-like green.
> The suite is decorated in classic white and yellow.
> The suite has a seperated living room.

> 斯特拉文斯基套房以作曲家斯特拉文斯基的名字命名，他的音乐常以春天为背景，因此，室内装饰以象征春天的绿色等欢乐的色彩为主色调。
> 套房以经典的白色和黄色为主色调。
> 这间套房拥有一间独立的客厅。

> The Mariinsky Suite is decorated in light blue tones to match those of the interior of the Mariinsky Theatre, and has a theatrical ambience to it.
> The Amber Suite is named in honour of the famous Amber room at the Catherine Palace in Tsarskoye Selo, and it features warm amber tones.
> In this green Stravinsky Suite, there are red furnitures.

> 马林斯基套房采用淡蓝色调装饰，与马林斯基剧院的室内装饰相呼应，并拥有剧场般的氛围。
> 琥珀套房以沙皇村凯萨琳宫中著名的琥珀室命名。琥珀套房以温馨的琥珀色为主色调。
> 绿色的斯特拉文斯基套房采用红色家具作装饰。

> The lounge is separated with the bedroom.
> The Faberge Suite is named in honour of the Russian jeweller, Carl Faberge. The interior is designed in the finest traditions, embodying his works of art.
> The colour scheme is centred around shades of pink, lilac and golden tones, and the suite is furnished with light, almost white coloured furniture encrusted with precious stones and patina.

> 休息室与卧室分离。
> 费伯奇套房以俄罗斯珠宝商卡尔·费伯奇的名字命名。室内采用最经典的传统风格设计，充分展示了他的艺术杰作。
> 室内装饰主要以粉色、淡紫色和金色为主色调，套房中的家具采用淡雅色调或几乎为白色，家具上饰有铜绿或珍贵的宝石。

> The blue bedsheet is striking in this suite.
> The Romanov Suite is named in honour of the Imperial Russian Dynasty, members of which regularly frequented the hotel.
> This suite has a truly palatial atmosphere and is furnished with antique furniture featuring decorative gold moulding.
> The suite stunned contemporaries with the unprecedented opulence of its interiors.

> 套房中的蓝色床单显得格外耀眼。
> 罗曼诺夫套房以俄罗斯王朝的名字命名，当时王朝官员经常下榻该酒店。
> 这间套房拥有纯正的宫殿般的氛围，采用带有金色镶边的古典家具作装饰。
> 这间套房以其无比奢华的室内装饰，让世人震惊。

HOTEL PRINCIPE DI SAVOIA

普林西比塞维亚酒店

Completion date（项目建成时间）：2009 (renovation) / Location（项目地点）：Milan,
Italy / Designer（设计师）：Thierry Despont, Francesca Basu / Photographer
（摄影师）：Hotel Principe di Savoia / Area（室内面积）：41,000 sqm /
Renovation area（翻新面积）：5,460 sqm

Hotel Principe di Savoia Milano is in the central business district in Milan, Italy. It is just 100 metres from Milan's Repubblica Metro.

Renowned New York–based designer, Thierry Despont is responsible for the redesign of the public areas of the hotel. The lobby is now a much lighter, more welcoming entrance to this prestigious palace hotel. On arrival, the Il Salotto Lobby Lounge on the right–hand side is a convenient meeting space for the well-heeled Milanese and guests alike to enjoy an aperitif surrounded by sumptuous Italian furniture. To create an immediate impact on arriving at the Principe di Savoia, Despont has added classical paintings by famous artists such as Luca Giordano and a spectacular custom–made Murano glass chandelier created by Barovier and Toso. This total transformation creates the appropriate feeling of grandeur and occasion for those arriving at this iconic hotel.

Thierry's vision is also to bring the Principe Bar to life making it more vibrant and engaging, while retaining some of the original features such as the beautiful marble and wood panelling. Despont has managed to create a perfect balance between the classical and innovative styles that embody the Principe di Savoia. The centrepiece of the room is a custom–made banquette that "wraps around" a grand piano, perfect for intimate musical soirees; a further reason why this Bar is sure to continue to be the place to be and be seen in Milan!

The redesign of the Imperial Suite was overseen by Celeste dell'Anna and combines a series of contemporary and classical elements. The 230 square metres suite has a large sitting room, featuring a handcrafted mini bar console and crocodile skin writing desk. Striking paintings featuring interpretations of contemporary masterpieces have been specially created by Celeste dell'Anna. The bedroom boasts a four–poster bed in the richest fabrics and a large walk–in wardrobe capable of housing all the latest purchases from the nearby shopping district.

FORMA EIUS IN SPECIE ORBIS ABSOLUTI GLOBATA
NON SOLO QUIA TALIS FIGURA ORBIS

> The veranda of the meeting room has a view of the charming garden.
> 会议室的过道可一览花园的迷人景色。

普林西比塞维亚酒店位于意大利米兰的中央商业区，距离米兰的雷里布里卡地铁站仅100米远。

著名的纽约设计师蒂埃里·德仕庞负责酒店公共区的翻新设计。现如今的酒店大堂更加明亮，更加温馨，作为这家享有声望的皇宫酒店的入口，给顾客一种宾至如归的感觉。一进入酒店，游客就会被右手边的Il Salotto大堂休息室所吸引，它是品位高雅的米兰人和游客们方便的聚会场所，他们可在摆放着豪华意式家具的背景下品尝一杯开胃酒。设计师在入口处增加了卢卡·焦尔达诺等著名艺术家的经典画作，并专门定制了由Barovier & Toso公司制作的慕拉诺水晶吊灯。这种整体性的转变塑造了一个庄严宏伟的酒店形象，使游客一进入这家标志性的酒店，便感受到它的庄重与辉煌。

蒂埃里·德仕庞还试图赋予普林西比酒吧以强大的生命力，使它焕发出勃勃生机，从而更具吸引力，同时保存酒店的部分经典特色，比如漂亮的大理石和木板等。德仕庞成功地打造了普林西比塞维亚酒店特有的创新风格与古典风格间的平衡。室内的核心是一条量身打造的人行道，人行道环绕着一架巨大的钢琴，这里是举办亲密音乐会的理想场地，更使普林西比酒吧成为米兰人聚会与交流的绝佳场所。

室内设计师Celeste dell'Anna负责皇室套房的翻新设计，设计过程将现代与古典元素相结合。230平方米的套房拥有一间宽敞的客厅，客厅内有一个手工制成的迷你吧台和一张鳄鱼皮的办公桌。设计师还特殊放置了一些现代主义风格的经典壁画。

卧室内摆放着一张四柱床，床上摆着珍贵的床品，还有一个内嵌式衣柜，可供客人摆放从附近商场购买的所有新款衣物。

> The entrance is unforgetable with its wooden decoration and grand ceiling.
> 酒店入口的木质装饰和华丽的屋顶给人留下深刻印象。

> Presidential Suite plan
> 总统套房平面图

1. Sauna	1. 桑拿房
2. Turkish bath	2. 蒸汽浴室
3. Tub	3. 浴盆
4. Salone	4. 客厅
5. Bathroom	5. 浴室
6. Dining room	6. 餐厅
7. Kitchen	7. 厨房
8. Bedroom	8. 卧室

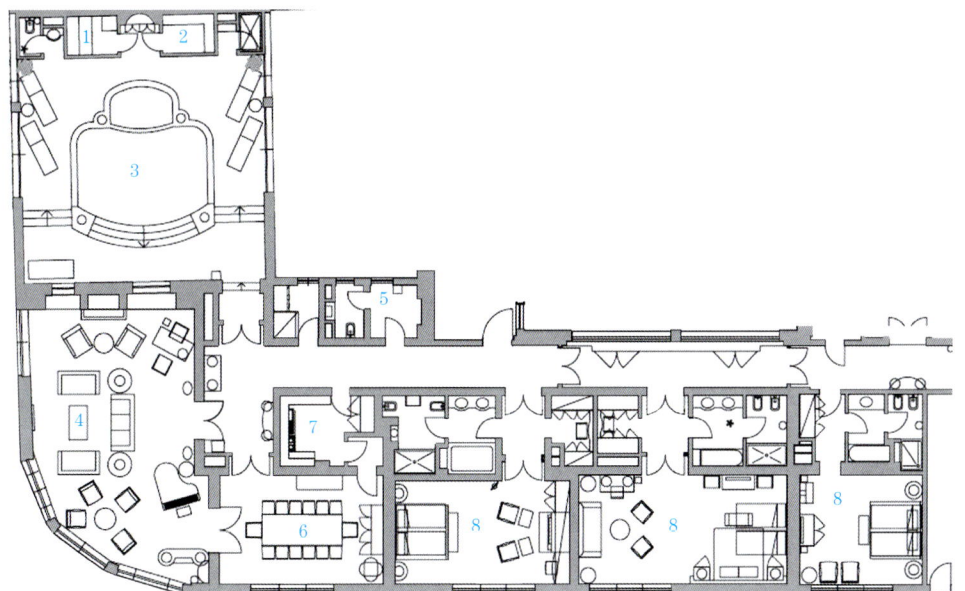

> The Imperial Suite is an irresistible luxury flat with colourful ornaments.

> Freshly redecorated Ambassador Suites are a symphony of unique details.

> The Principe Suite is enriched by precious and colourful fabrics and furniture that recall the classic heritage.

> The walls are decorated with paintings, reinterpreting the most famous masterpieces of the 20th century.

> 皇室套房中色彩斑斓的装饰将其打造成无比奢华的公寓。

> 刚刚翻新的大使套房十分注重细节装饰，别致典雅。

> 珍贵而又多彩的织物与家具将普林西比套房装点得格外华丽，是经典的传承。

> 墙面上挂着的壁画是对20世纪最著名的杰出画作的重新诠释。

> The decoration of Ambassador Suite is rich and of extreme luxury, created in the trendiest tones of cream and violet.

> 大使套房采用最时尚的奶白色和紫罗兰色，富贵典雅、尽显奢华。

> The Royal Suite features a spacious hall with smart polychromatic marble floor, and sitting room with rich silk panelling.

> 皇室套房拥有一个宽敞的大厅，大厅地面采用亮丽多彩的大理石地板，客厅墙面采用华丽的丝绸作装饰。

> Imperial Suite plan

> 皇室套房平面图

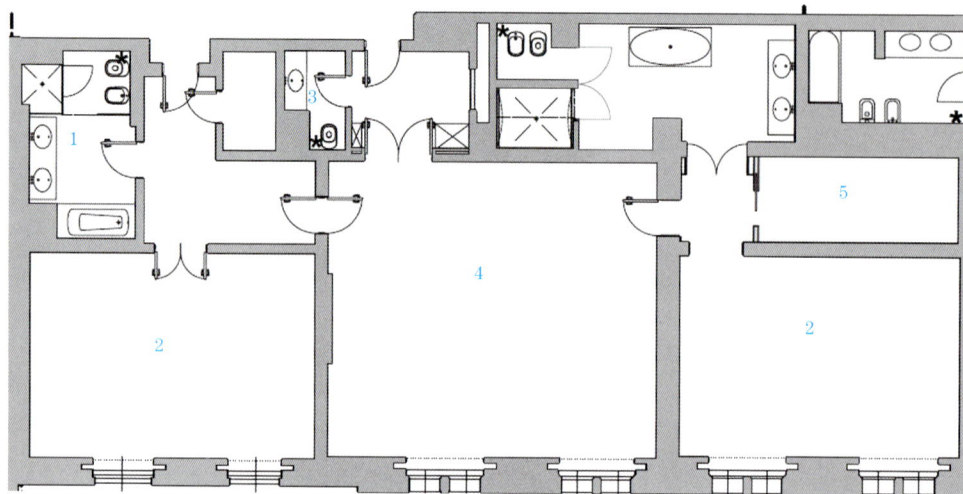

1. Bathroom	1. 浴室
2. Room	2. 卧室
3. Toilet	3. 卫生间
4. Lounge	4. 休息室
5. Dressing room	5. 化妆室

> Presidental Suite features empire style, original antique furniture and an elegant working fireplace in the living room.

> 总统套房以帝国风格为特色，采用原创的古典家具，客厅内正在燃烧的壁炉更烘托出优雅的氛围。

> Principe Suite plan

> 普林西比套房平面图

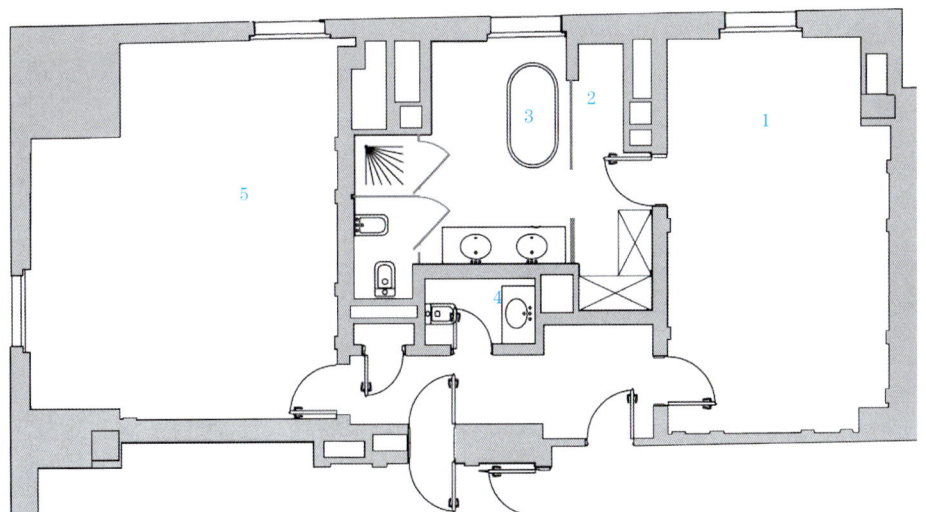

1. Room   1. 卧室
2. Dressing room 2. 化妆室
3. Bathroom   3. 浴室
4. Toilet   4. 卫生间
5. Lounge   5. 休息室

> The bedroom of Presidencial Suite contrasts two key colours.
> Located on the first floor, the prestigious Royal Suite is one of the signature suites of Hotel Principe di Savoia.
> Chandeliers, marble and bronze wall lamps, and fine objects in marble and granite complete the luxurious location.
> The soft natural colours, contrasting the warm tone of brown, violet and beige, offer exclusive relax and comfort.

> 总统套房的卧室中，两种主色调形成鲜明的对比。
> 著名的皇室套房位于一楼，是普林西比塞维亚酒店的招牌客房之一。
> 枝形吊灯、大理石与青铜台灯，还有花岗岩材质的精致物品打造了无比奢华的布置。
> 自然、柔和的色调与棕色、紫罗兰色和米黄色的温暖色调形成鲜明对比，带给游客无比放松与舒适的环境。

> The soft natural colours of the walls perfectly harmonise with the rich Italian furniture.

> 墙面柔和、自然的色调与华丽的意大利家具完美呼应。

> The walls claim the noblest reinterpretation of classical boiseries and fabrics, and the style captures the modern but elegant trend.
> 墙面是对古典的细木护壁板和织物最华丽的诠释，室内装饰风格将现代的时尚与古典的优雅相融合。

LE MEURICE

莫里斯酒店

Completion date (项目建成时间) : 2007 / Location (项目地点) : Paris, France /
Designer (设计师) : Philippe Starck / Photographer (摄影师) :
Le Meurice / Area (室内面积) : 15,200 sqm

Le Meurice is a five-star hotel in Paris, located opposite the Tuileries Garden, between Place de la Concorde and the Musée du Louvre.

The quintessential rendez-vous for discerning Parisians, Bar 228 offers a warm and sophisticated environment, reinterpreted by Philippe Starck in December 2007. With soft, sumptuous furnishings and details, and comfortable seating throughout, it's the perfect place to unwind and enjoy delicate cocktails, such as the "Starcky" or "228".

Located on the mezzanine level, The Spa Valmont for Le Meurice is a tranquil idyll with natural accents in marble, wood, stone and glass. To offer all clientele the luxury of highly efficient skin-care and the very best anti-aging treatments, they chose leading cosmetics brand, Valmont.

At Le Meurice, opulence and comfort are not confined to the hotel's public areas alone. Every room and suite is a spacious, peaceful oasis of calm, providing guests with the ultimate in luxury. Each of the seven floors has a distinct style, with 160 beautifully appointed rooms decorated in a style redolent of Louis XVI. All rooms feature exquisite marble bathrooms with shower and bath, with complimentary bathrobes and slippers. There are 45 suites and junior suites, most of which overlook the Tuileries Garden and have spectacular views of the Paris skyline. On the first floor, two Presidential Suites offer high ceilings and sumptuous furnishing throughout. Prestige Suites - on the second and third floors of the hotel, have their own beautifully designed living rooms.

True to the spirit of Le Meurice, modern comfort blends harmoniously with the charm of classic French furnishings in Louis XVI style. Period furniture is complemented by elegant fabrics, capturing the essence of Parisian luxury in a suite that is bathed in light. In addition to a courtyard view, it has a private entrance, a separate and spacious sitting room, a large writing desk, dressing room and mini bar. The exquisite Italian marble bathroom has both a bath and a walk-in shower. Time is suspended amid the tranquil atmosphere and elegantly muted palette of this suite.

> The lounge of lobby is full of golden decoration.
> 酒店大堂的休息室中布满了金色的装饰。

莫里斯酒店是巴黎一家五星级酒店，该酒店位于杜伊勒里花园的对面、巴黎协和广场与卢浮宫博物馆中间。

作为品位高雅的巴黎人的激情密约之地，228酒吧拥有温馨、高雅的环境，2007年由设计师菲力浦·斯塔克重新设计。采用柔和、豪华的家具、精致的细节设计与舒适的座椅，228酒吧是宾客放松身心的好地方。您可在此感受"Starcky"和"228"等鸡尾酒的浓郁幽香。

法尔曼SPA位于莫里斯酒店的夹楼中间，采用大理石、木材、石头和玻璃等原材料装饰，仿佛是一个幽静的田园。为提供给顾客高效的皮肤护理和最佳的抗衰老理疗，酒店选择上等的化妆品品牌——法尔曼。

在莫里斯酒店，不仅公共区域彰显着奢华与舒适，每一间客房与套房都宽敞明亮，仿佛是一个宁静的绿洲，游客可在此体验极度奢华。酒店共7层，每一层都拥有别致的风格，160间客房精美典雅，全都采用路易十六时代的风格设计。所有客房都配备精致的大理石浴室，浴室内配备淋浴器、浴盆，并贴心地准备了浴袍和拖鞋。酒店共有45间套房与标准套房，大部分客房内都可一览杜伊勒里花园的美丽景色，并可环视巴黎整个城市的夜景。一层的两间总统套房拥有高高的天花板，并且所有家具都无比奢华。酒店二层与三层的豪华套房拥有独立的客厅，客厅采用精美设计。现代舒适与路易十六风格的法式古典家具完美融合，每一处设计都体现了莫里斯酒店的精神。套房沐浴在灯光之下，套房内的仿古家具在优雅的织物的映衬下显得更加别致，捕捉住了巴黎式奢华的精髓。套房内除了可以欣赏美丽的庭院景观外，还拥有一个私人入口、一个独立的宽敞客厅、一个大型的写字台、一个化妆间和一个迷你吧台。精致的大理石浴室内不仅配备了一个浴盆还安装了一个步入式淋浴房。柔和的色调与静谧、优雅的氛围使时光在套房内停驻。

> The open reception area affords views of the hotel's three main public spaces: the Restaurant Le Meurice, the Restaurant Le Dali and the Bar 228.

> 开放的接待区可一览酒店三个主要公共区的景色：莫里斯餐厅、达利餐厅和228酒吧。

> The lobby is definitely Dali-esque, and there are many design elements created by Philippe Starck.

> 酒店大堂完全采用达利派风格设计，并且融合了许多菲力浦·斯塔克的设计元素。

> Expansive lobby greets guests with a divine white and green marble floor and an exquisite collection of antiques and artwork.

> 广阔的酒店大堂采用高贵的白绿色大理石地板和精致的古董与手工艺品作装饰，向广大游客发出盛情邀请。

> Belle Etoile Royal Suite plan

> 百丽多华套房平面图

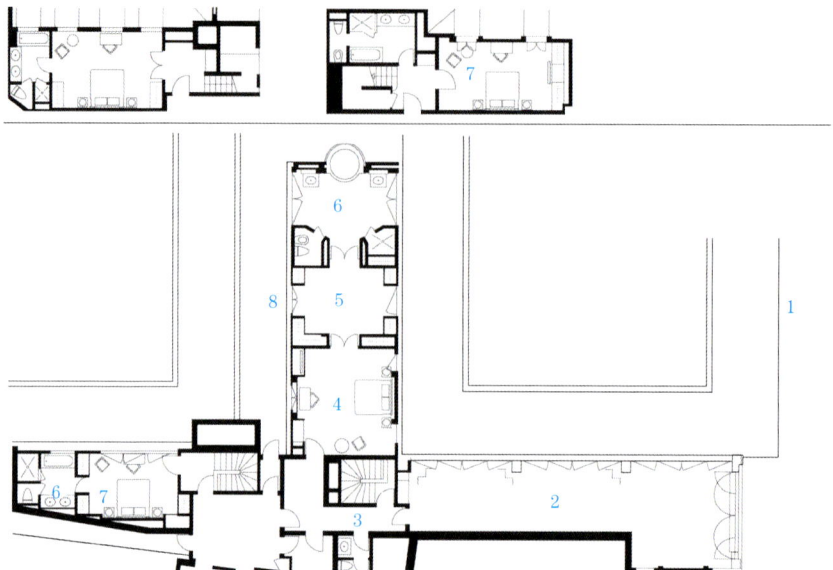

1. Tuilleries garden view   1. Tuilleries花园景观
2. Sitting area   2. 休息区
3. Entrance   3. 入口
4. Master bedroom   4. 主卧室
5. Dressing room   5. 化妆室
6. Bathroom   6. 浴室
7. Bedroom   7. 卧室
8. Terrace   8. 露台

> The warm and hushed decor remains faithful to its origins with wooden bar 228 stools and leather chairs.
> 木质的228酒吧高脚凳和皮革座椅，温暖而宁静的装饰是对古典装饰的经典演绎。

> The sitting room of Belle Etoile Royal Suite is decorated in a nineteenth-century style.
> The sitting room of this suite is in cool colour.
> Decorated in the French classical tradition with Louis XVI style furniture, this suite was the backdrop for the artist's exploits.

> 百丽多华皇室套房的客厅采用19世纪的装饰风格装饰。

> 套房客厅采用冷色调装饰。

> 这间套房采用传统的法式经典风格装饰，搭配路易十六时期的家具，是艺术家创作的天堂。

> Period furniture is complemented by elegant fabrics, capturing the essence of Parisian luxury in a suite that is bathed in light.
> Luminosity and a unique sense of tranquillity fill this Executive Room, which overlooks the courtyard.
> The master bedroom of Belle Etoile Royal Suite features a spacious bed with a midnight-blue canopy.
> The bedroom of Presidential Suite experiences the magic and splendour of the Versailles style, with towering moulded ceilings.

> 仿古家具搭配优雅的织物，沐浴在阳光中的套房凝聚了巴黎奢华的精髓。
> 行政套房俯视庭院，客房宽敞明亮、十分安静。
> 百丽多华皇室套房的主卧室中摆放着一张宽敞的卧床，卧床上方挂着一个午夜蓝调遮篷。
> 总统套房的卧室采用高耸的定制屋顶，彰显了凡尔赛风格的大气与辉煌。

157

> Every modern comfortable bedroom has been seamlessly incorporated into the charm of classic French style.
> The Executive Junior Suite, which overlooks the courtyard, has a private entrance and large writing desk.
> Time is suspended amid the tranquil atmosphere and elegantly muted palette of this Junior Suite.

> 卧室中，每一处彰显现代舒适的装饰都与法式经典风格的迷人魅力天衣无缝地融合在一起。

> 俯视庭院的标准行政客房拥有一个私人入口和一个大大的写字台。

> 标准套房拥有柔和的色调和静谧的氛围，时光在这里停驻。

VENEZIA TOWER AT VENETIAN HOTEL & RESORT

威尼斯人度假村酒店中的威尼斯酒店

Completion date（项目建成时间）: 2007 / Location（项目地点）: Las Vegas,
United States / Designer（设计师）: SFA Design / Photographer（摄影师）:
Erhard Pheiffer / Area（室内面积）: 8,250 sqm

Located in the heart of the Las Vegas Strip, the new Venezia Hotel Tower is comprised of an additional 1,013 luxury suites and featuring a lavish pool deck.

For the Venezia Tower, SFA Design created a hideaway within the hotel that builds upon the Venetian's existing Italian Renaissance theme, while evoking the ambiance of a private palazzo. To ensure a more intimate, "boutique"-style hotel-within-a-hotel, SFA designed private seating areas, using soft lighting and deep woods that offer an air of welcome, while grand fixtures, luxe fabrics and rich, intricate patterns on the flooring and furniture evoke the ambiance of an elite estate.

Venetian's Italian Renaissance philosophy works with the scale and ambiance of a private palazzo. SFA meticulously crafted the reception areas, featuring a dramatic rotunda with original murals reminiscent of the frescoes of early 18th century grottos. Each layer of the dramatic covered ceiling, although separate, is visually cohesive with romantic references to Neptune, and his gift to Venice. Intricate stone patterns, rubbed plaster walls, and custom–designed light fixtures reinforce the formality of the architecture. The ancillary areas are left refreshingly devoid of superfluous furnishings to capture the spirit of the timeless edifices of Italy. Each transition extending from the main rotunda has been designed to enhance the guest experience.

The extraordinarily large guest suites (known as the largest guest suites in Vegas!) have been refreshed from the main hotel's original design. Each room, a private sanctuary with its own art collection, is furnished in a relaxing combination of old world luxury and ultramodern styling. A reflection of 21st century sophistication, the generously appointed five-fixture bath is finished in gleaming, polished marble and gold fixtures. The separate sleeping and living areas frame the dramatic views of the infamous strip below. The Venetian's award-winning suites, the standard luxury suite accommodations of Venezia Tower include elegant, modern Italian décor and average over 65 square metres, but their three metres ceilings allow for a greater sense of space.

威尼斯酒店地处拉斯维加斯大道的中心位置,除了主要的酒店设施外,还拥有1,013个奢华的套房和一个浪漫的特色泳池甲板。

SFA Design 设计事务所在原意式文艺复兴的酒店主题基础上又打造了一个世外桃源,从而营造了一种私人豪华宫殿的氛围。为打造一个酒店中的时尚酒店,形成更亲密的氛围,SFA Design 设计事务所设计了一些私人座位区,并使用柔和的灯光与暗色的木材,给游客一种宾至如归的感觉,而豪华的布置、奢华的织物与地板上彰显富贵的错综图案与家具则打造了经典豪宅的奢华氛围。

威尼斯酒店的意大利文艺复兴哲学与私人宫殿的规模和氛围相匹配。SFA Design 设计事务所精心设计

了接待区,接待区主要突出一个圆形大厅,大厅用古典壁画作装饰,富丽堂皇,壁画的内容酷似18世纪早期的洞穴。引人注目的天花板的每一层都各自独立,然而又彼此相连,上面带有象征海神的图案及海神带给威尼斯的礼物。复杂的石头花纹、打磨的灰泥墙、定制的灯具都增加了建筑的庄严感。配套的服务区没有采用奢华的装饰,给人一种清新自然的感觉,体现了意大利建筑的永盛不衰。从圆形大厅向外的每一种过渡都丰富着游客的感官体验。

无比宽阔的大型套房(据说是威尼斯最大的客房)经翻新后,每一间套房的装饰都融合古典奢华与超现代设计于一身,仿佛是收藏个人艺术品的神殿。豪华的五件式浴房采用打光的大理石和黄金部件

精制而成,反映了21世纪的精湛技艺。独立的卧室与客厅可直观下层朴素的带状空间。威尼斯酒店的套房曾获大奖,标准的奢华套房突出优雅的现代装饰,套房平均面积为65平方米,而天花板高达3米,呈现出更大的空间感。

163

> The lobby lounge is decorated with golden fabric and incredible ceiling.

> 大堂休息室采用金色织物作装饰，并拥有华丽壮观的屋顶。

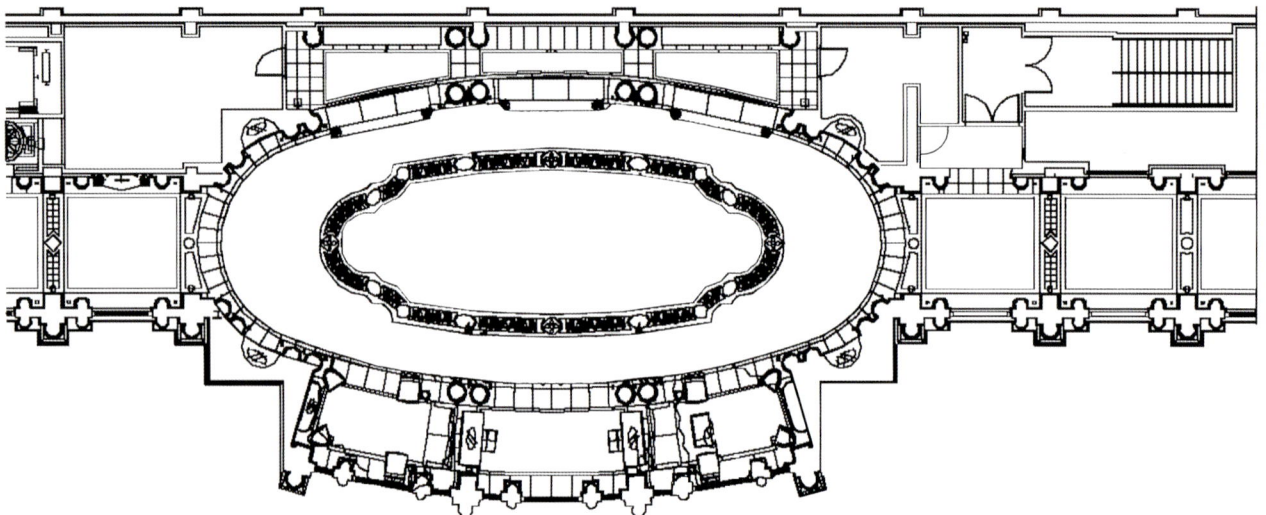

> Lobby fixture furniture plan 1

> 大堂家具布置平面图1

> The black table in the club lounge is unique.
> 会所休息室中的黑色桌子十分别致。

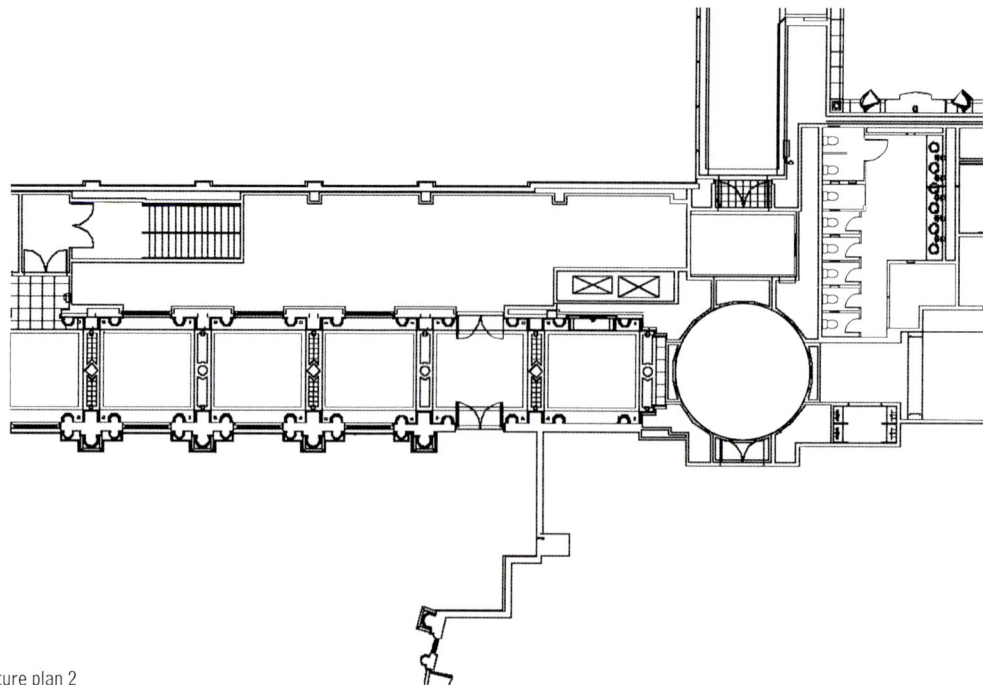

> Lobby fixture furniture plan 2
> 大堂家具布置平面图2

> The elevator lobby has wooden wall and bright light.
> 电梯门厅中拥有木质墙壁和明亮的灯光。

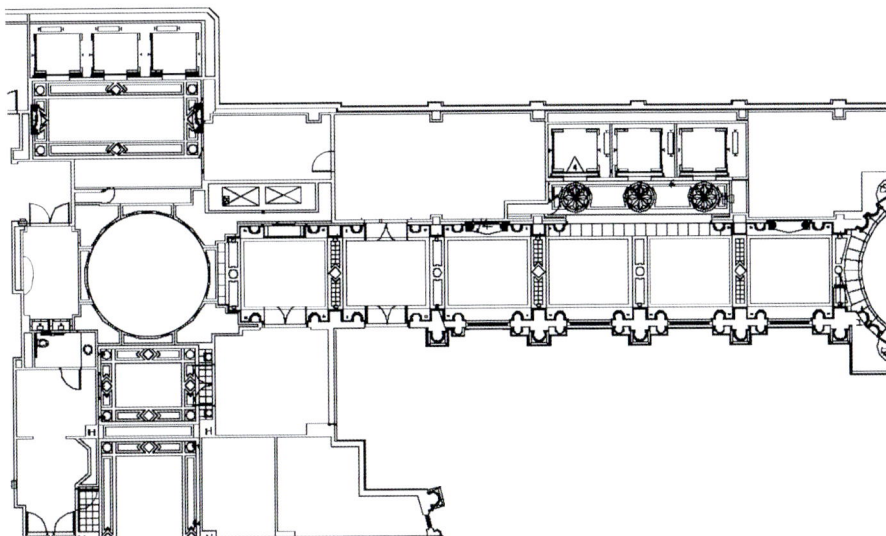

> Lobby fixture furniture plan 3
> 大堂家具布置平面图3

THE RITZ LONDON

伦敦里兹酒店

Completion date (项目建成时间) : 2003 / Location (项目地点) : London, UK
Designer (设计师) : Charles Mewès, Arthur Davis / Photographer (摄影师) :
LHW / Area (室内面积) : 6,100 sqm

This exclusive city hotel lies at the heart of London and is the ideal starting point from whence to explore the English metropolis' top sights.

The actual design of the hotel is the work of architects Charles Mewès and Arthur Davis. Mewès, a Frenchman and London–born Davis had worked with César Ritz before on the Hotel Ritz in Paris, and The Carlton in London. For The Ritz London they drew up a stunning French chateau-style masterpiece with a wealth of clever details: light wells allowed rooms with no outside windows natural light, projecting dormer windows and tall chimneys broke the skyline.

Leading off the lobby is the Long Gallery, a vaulted space that runs almost the whole length of the building. The absence of walls or doors means it's possible to see down its whole length through The Restaurant and out over the hotel's Italian Gardens and across Green Park. Off the Long Gallery are many of the hotel's key rooms, all with their own fascinating histories: The Restaurant, often described as one of the most beautiful dining rooms in Europe features so many chandeliers on the ceiling, which had to be specially reinforced to cope with their weight.

To stay at The Ritz is to enjoy the ultimate in style, service and sophistication. They offer five types of suites, all with beautiful, fully restored period interiors and most with the facility to connect further rooms to each suite.

Some of the details are less functional– the copper lions on the corner of the roof are purely decorative. Inside, the French theme continues. Mewès designed the interiors with a single Louis XVI theme incorporating all of César Ritz's many requirements like double glazing, a sophisticated (for the time) ventilation system and a bathroom for every guestroom. During the Second World War, The Marie Antoinette Suite, a smaller private dining room was used as a venue for Summit Meetings by Churchill, de Gaulle and Eisenhower.

> The William Kent Banqueting Room is dominated by a breathtaking coved ceiling in grand Italian Renaissance style.
> The Palm Court epitomises the elegantly frivolous comfort of Edwardian high life.

> 威廉·肯特宴会厅以壮观的拱形屋顶为特色，屋顶采用意大利文艺复兴的风格设计。
> 棕榈厅是爱德华时期高雅、舒适生活方式的缩影。

这家顶级的城市酒店坐落于伦敦市中心，是探索英国大都市绝美景观的理想出发地。

建筑师查尔斯·缪伊斯和亚瑟·戴维斯负责酒店的具体设计。法国设计师缪伊斯、伦敦出生的设计师戴维斯与酒店主人凯撒·里兹共同合作完成了伦敦里兹酒店的设计，之后两位设计师又共同设计了巴黎的里兹酒店和伦敦的卡尔顿酒店。他们为里兹酒店拟定了神奇的城堡式风格，设计同时涵盖了诸多巧妙的细节，例如，部分房间没有安装外部窗户，而天井的设计则保证了其室内的自然采光；突出的屋顶窗和高高的烟囱打破了酒店上方的天际线。

酒店大堂与长廊相连，长廊是一个长长的拱形空间，贯穿了整栋建筑的横向空间。长廊中间取消了任何门与墙面的设计，因此，游客的视线可以沿整个长廊一直漫延至The Restaurant餐厅，甚至将室外的意大利花园囊括其中，并且穿越整个绿色公园。长廊远处是酒店的主要房间，所有房间都有着神奇的历史——常被人描述为欧洲最美餐厅之一的The Restaurant 餐厅中，最突出的特点就是天花板上挂着的诸多枝形吊灯，天花板经过特殊的加固设计以承担吊灯的重量。

下榻伦敦里兹酒店，您将体验到美妙绝伦的酒店风格、无比贴心的服务和精湛的技艺。该酒店共有5个类型的套房，所有套房都装修精美，重现了古典的室内氛围。每间套房都配备完善的设施，这些设施将各个套房相连。

一些细节设计相对功能性而言则更多地发挥了装饰性，比如屋顶角落里的铜狮子则纯粹是一种装饰。室内设计主要突出法式经典主题。缪伊斯将室内空间设计成了单一的路易十六主题，同时结合了凯撒·里兹的许多要求，如双层玻璃、高级的通风系统和每间客房内配备的浴室等。现在的玛丽·安托瓦内特套房在第二次世界大战期间曾是一间小型的私人餐厅，当时常是邱吉尔、戴高乐和艾森豪威尔等首脑召开政府高层会议的地方。

> Leading off the lobby is the Long Gallery, a vaulted space that runs almost the whole length of the building.
> 连接酒店大堂的长廊是一个贯穿整个酒店横向空间的拱形空间。

> With its opulent gold leaf mouldings, slender Adamesque pilasters and shallow-arched ceiling, this stately room epitomises the grand and the graceful style of the house.

> 庄严的大厅采用华丽的金叶线条、纤细的亚当式栏杆和拱形屋顶作装饰，体现了酒店辉煌、高雅的设计风格。

> Ground floor plan

> 一层平面图

1. Reception
2. The Music Room
3. Hall
4. The Queen Elizabeth Room
5. The Wimborne Room
6. The William Kent Room
7. Bathroom

1. 前台
2. 音乐室
3. 大厅
4. 伊莉莎白女王宴会厅
5. 温伯恩宴会厅
6. 威廉·肯特宴会厅
7. 浴室

> Beneath a gilded ceiling hung with a crystal chandelier and yellow silken damask walls, the beautiful inlaid mahogany table can accommodated up to 16 guests.

> The vast floor to ceiling windows, the rich and varied use of soft pink, pale green and the dazzling garlands of chandeliers are reflected in the wall of panelled mirrors.

> The Prince of Wales Suite is a beautiful apartment—style penthouse, which includes its own drawing room.

> 镀金的屋顶上悬挂着一盏亮丽的水晶吊灯，四周是黄色的丝缎墙壁，精美的嵌花红木桌子可供16名客人同时用餐。

> 镶边的镜面墙壁映照着巨大的落地窗、淡粉色和淡绿色的奢华装饰，和水晶吊灯周围耀眼的花环。

> 威尔士亲王套房拥有独立的会客厅，采用漂亮的顶层公寓风格设计。

> The 40-capacity Queen Elizabeth Room has quite a different character with its blue-patterned wallpaper and dainty sofas, and it could be the drawing room of a stately home.

> The oval master bedroom of Royal Suite has great view over Green Park.

> Unique furnishings and beautifully maintained décor create a very private space within the hotel.

> With marble halls in many, deep carpets, rich, heavy curtains, antique paintings and furnishings, breathtaking might be a better adjective.

> 可容纳40人的伊莉白女王休息室呈现出截然不同的景向——蓝色印花的壁纸和精美的沙发，很像是豪华古宅的会客厅。

> 皇室套房中椭圆形的主卧室中可一览绿色花园的美丽景色。

> 别致的家具与布置精美的装饰品打造了酒店内的隐蔽空间。

> 采用多层深色地毯、华丽的帷幕、古典的壁画和家具装饰的大理石大厅，用"令人叹为观止"来形容似乎更贴切。

FAIRMONT HOTEL VIER JAHRESZEITEN

费尔蒙特维耶亚利策芬酒店

Completion date（项目建成时间）: 2010 / Location（项目地点）: Hamburg, Germany /
Designer（设计师）: Tillmann Wagenknecht / Photographer（摄影师）:
Fairmont Hotels&Resorts / Area（室内面积）: 19,000 sqm

Prominently located on the western side of the Inner Alster Lake shore in Germany, The Fairmont Hotel Vier Jahreszeiten lies in the heart of the prime retail and commercial district of Germany.

Luxury Vier Jahreszeiten Hotel in Hamburg features the stylish entrance hall and check–in desks with rich ornamented marble floor tiles and wooden wall coverings. The halls and parlors are one and all equipped with modern technology: both the ceiling projectors and screens come out of LED lighting technology, giving it almost any desired colour, which is controlled by touch screens exactly (and surprisingly easy to use).

The stylish restaurant or dinning room is decorated with rich ornamented marble floor tiles, wooden wall coverings and rich candeliers. Located on the ground floor, The Wohnhalle is the heart of the Fairmont Hotel Vier Jahreszeiten, and inspires a turn of the century country house atmosphere. A wood burning fireplace adds the cosy oasis to calmness of this living room lounge. The lounge features wood wainscotting, marble floors, oil paintings, stuccoed ceilings, lavish red drapes and antique armchairs and sofas. The menu includes light snacks, coffee, cake specialties and a classic British afternoon tea.

The royal guest apartment of the luxury Vier Jahreszeiten Hotel in Hamburg is set with stylish old–fashioned design and a massive comfortable wooden bed with wooden canopy. The stylish guest bedroom is decorated with contemporary styled bedroom furniture design with comfortable bed and fitted wardrobe with mirror doors. All bathrooms have been duped by White Etage, 15 rooms, of which "Sir Peter Ustinov Suite" have been upgraded and refurbished.

Celebrity Suites offer a breathtaking view of the Inner Alster Lake. They are extremely luxurious and spacious, featuring a hardwooden foyer, an elegant living room, perfect for entertaining and a master bedroom with a king–size bed. Living room and bedroom are separated by a door for privacy. These large Lakefront Suites are furnished with valuable antiques, original paintings and beautiful mirrors. The Master Bathroom offers a separate glass shower stall, a soaking tub, double basins, a bidet and a heated floor.

> The entrance of Haerlin Restaurant is decorated with wood.

> 哈尔林餐厅的入口采用木材作装饰。

> The entrance of Fairmont Hotel Vier Jahreszeiten is magnificent, with the red carpet on the ladder.

> 费尔蒙特维耶亚利策芬酒店入口的楼梯上方铺设红色地毯，十分华丽。

费尔蒙特维耶亚利策芬酒店位于德国内阿尔斯特湖西岸，地处德国主铁路沿线和商业区的核心位置。奢华的汉堡维耶亚利策芬酒店拥有时尚的入口大厅和前台，内部装饰高贵典雅的大理石地砖和木制的墙面材料。大厅与会客室合二为一，采用现代化设施：云幕灯和显示屏都采用发光二极管照明技术，完全采用触摸屏控制，可以呈现出任何想要的颜色。

时尚的餐厅采用富丽堂皇的大理石地砖、木质的墙面材料和奢华的枝形吊灯作装饰。Wohnhalle餐厅位于该酒店的中心，打造了一种19世纪末20世纪初乡村房屋的氛围。燃木壁炉为客厅休息室的宁静带

来了一片惬意的绿洲。休息室采用木质的护墙板材料、大理石地面、油画、用灰泥粉饰过的天花板、大红色的窗帘、古色古香的座椅和沙发。餐厅提供一些小点心、咖啡、各式特色的蛋糕和经典的英式下午茶。

奢华的汉堡维耶亚利策芬酒店的皇家公寓客房采用高雅的古典式设计，室内配备一张宽大的、带木质遮篷的舒适木床。时尚的卧室采用现代风格的卧室家具——舒适的床与带玻璃门的壁柜。White Etage的15间客房都配备浴室，其中，Sir Peter Ustinov套房曾被翻新。

"名流"客房内可一览内阿尔斯特湖的壮观景色。

客房宽敞明亮、装饰典雅，包括一个硬木的门厅、一间优雅的客厅，客厅是举行娱乐活动的理想场所，还有一间配备特大号床的主卧室。客厅与卧室中间用一扇门隔开，以保护客人隐私。这些大型的湖畔套房都采用价格高昂的珍贵古董、原创的图画作品和亮丽的镜子作装饰。主卧室内还配备一个独立的玻璃浴室、一个浸泡式浴缸、一个双人浴盆、一个坐浴盆和加热地板。

> The Lounge, at the heart of the luxury hotel in Hamburg, creates a unique atmosphere of elegance and cosiness.

> 位于汉堡这家奢华酒店中心区域的大堂营造了一种舒适、优雅的别致氛围。

> The green chair is marked in this space.

> 绿色座椅形成了空间的亮点。

> Meeting floor plan
> 会议室平面图

1. Grand ballroom	1. 大宴会厅
2. Tapestry	2. 挂毯
3. Gesellschafts Salon	3. Gesellschafts沙龙
4. Jahreszeiten Salon	4. 维耶亚利策芬沙龙
5. Small ballroom	5. 小宴会厅
6. Oval room	6. 椭圆形会议室
7. Toilet	7. 卫生间
8. Haerlin Salon	8. Haerlin沙龙
9. Condiment	9. 调料间
10. Club	10. 俱乐部
11. Lobby	11. 大厅
12. Restaurant Haerlin	12. Haerlin餐厅

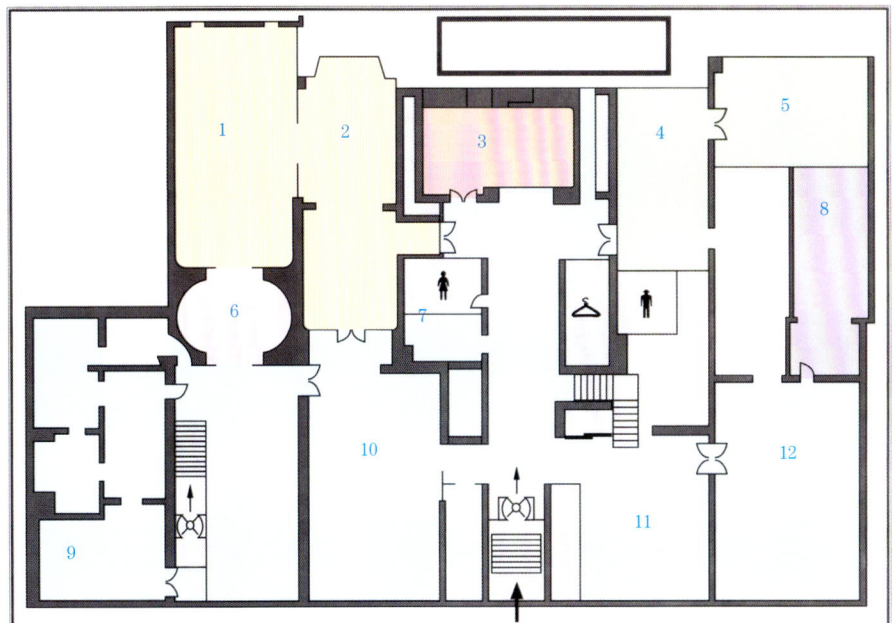

> The grand mahogany and marble lobby create an atmosphere much like that of a palatial estate.
> The restaurant features marble floors, oil paintings, lavish red drapes and antique armchairs.
> The small ballroom includes historic paintings and elegant curtain with beautiful architectural details.
> A wood burning fireplace adds the cosy oasis to the calmness of this living room lounge.

> 红木和大理石打造的酒店大堂营造出了宫殿般的奢华氛围。
> 餐厅以大理石地板、油画、奢华的红色窗帘和古典扶手椅为特色。
> 小型宴会厅采用古典壁画、精致的窗帘作装饰，并突出了优美的建筑细节设计。
> 燃木壁炉为宁静的客厅休息室带来了一片惬意的绿洲。

> In the tradition of the Biedermeier style, The Café Condi is especially well-known for the variety of its breakfast buffet.
> Haerlin Restaurant has oil paintings and green plants as décor.
> This small ballroom with its wooden fireplace has peaceful atmosphere.

> Condi餐厅采用传统的比德迈风格装饰，因其早餐自助菜肴的多样性而备受客人欢迎。
> 哈尔林餐厅采用油画和绿色植物作装饰。
> 采用木质壁炉装饰的小型宴会厅拥有宁静祥和的氛围。

> The comfortable Junior Suite is elegantly decorated usually with a sofa and two armchairs as well as a coffee table.

> Fairmont guest room is furnished with specially designed oversized queen-size bed with LED reading lamps.

> The Celebrity Suite is furnished with valuable antiques, original paintings and beautiful mirrors.

> 舒适的标准客房装饰优雅，常配备一张沙发、两把扶手椅和一张咖啡桌。

> 费尔蒙特客房采用特殊设计的超大号卧床和LED台灯作装饰。

> 名流套房采用昂贵的古玩、古典壁画和精美的镜子作装饰。

> With high ceilings, the Signature Room has a generous and exclusive atmosphere.
> Celebrity Suites offer breathtaking views of the Inner Alster Lake, extremely luxurious and spacious.
> The Exclusive Signature Suite is designed in alcove style and disposes of a size of 80 square metres.
> Deluxe Double Room is spacious and elegantly decorated, facing the luxurious, quiet courtyard.

> 招牌客房拥有高高的天花板，营造了大气与别致的室内氛围。
> 名流套房十分宽敞、奢华，并可一览内阿尔斯特湖的迷人景色。
> 别致的招牌套房采用壁龛风格设计，面积达80平方米。
> 豪华的双人客房宽敞明亮、装饰优雅，正对着奢华、安静的庭院。

FAIRMONT EMPRESS HOTEL

费尔蒙特皇后酒店

Completion date（项目建成时间）: 2010 / Location（项目地点）: Victoria, Canada /
Designer（设计师）: Forrest Perkins / Photographer（摄影师）: Fairmont Hotels
&Resorts / Area（室内面积）: 29,000 sqm

Located in central Victoria, The Fairmont Empress is near the airport and steps from Royal British Columbia Museum, British Columbia Parliament Building, and Pacific Undersea Gardens. Dining options at The Fairmont Empress include four restaurants. A bar/lounge is open for drinks. Room service is available 24 hours a day. Recreational amenities include an indoor pool, a children's pool, a health club, a spa tub, and a sauna. The property's full-service health spa has body treatments, massage/treatment rooms, facials, and beauty services. This Victoria property has 85,000 square metres of event space consisting of banquet facilities, conference/meeting rooms, a ballroom, and exhibit space.

The Fairmont Empress Hotel in Victoria, BC features 477 luxuriously appointed guest rooms and suites, with magnificent views of the city, courtyard or spectacular Inner Harbour. With a selection of unique guest room sizes and categories to suit every taste, this landmark Fairmont property is a tribute to classic elegance in accommodation.

All of Signature Rooms have been personally selected for their unique features and benefits. These premier guest rooms are chosen for their distinctive room layout and spacious size, views of the Inner Harbour and distinctive architecture. Each of spacious and luxurious Junior Suites offers an open-concept, and is elegantly furnished with a sofa bed and distinct sitting area. Each of spacious One-Bedroom Suites features two rooms - a bedroom and parlor with dividing door. The parlor room is lavishly furnished with a sofa bed, and some with ample space to accommodate a rollaway bed. Most these Victoria hotel suites feature one bathroom, while a few are equipped with an additional powder room. Each of the Two-Bedroom Suites features three rooms in total: two bedrooms and one parlor. The master bedroom is furnished with one king-size bed, the second bedroom with two queen-size beds, and the parlor with a sofa bed and ample space for a rollaway. Each of these Victoria hotel suites is equipped with a private bathroom and full amenities.

> The world renowned tea lobby of The Fairmont Empress has served England's most beloved ritual and Victoria's grandest tradition of afternoon tea to famed royalty.

> 费尔蒙特皇后酒店中世界著名的茶室将为皇室贵族提供英国最受欢迎的礼仪服务和维多利亚最辉煌的传统下午茶服务。

费尔蒙特皇后酒店地处维多利亚市中心，靠近机场，距离皇家不列颠哥伦比亚博物馆、皇家不列颠哥伦比亚国会大厦和太平洋海底公园仅几步之遥。

费尔蒙特皇后酒店拥有4间餐厅，其中一间餐厅专门提供饮品。酒店客房全天候开放。酒店配备室内游泳池、儿童游泳池、健身俱乐部、温泉浴盆和桑拿浴等设施。服务设施齐全的保健温泉浴场还配备美体护理、按摩室或治疗室、面部护理和美容等多项服务。维多利亚费尔蒙特皇后酒店拥有85,000平方米的活动场地，其中包括各种宴会设施、会议室、一个舞厅和展示厅。

坐落于BC省维多利亚市的费尔蒙特皇后酒店拥有477个装修奢华的客房与套房，每间客房和套房都可一览城市、庭院或内港的美丽景色。客房大小与类别都经过精心挑选与设计，费尔蒙特酒店的地标性建筑传承了古典住宿酒店的优雅。

酒店的所有招牌客房都采用个性化设计，具有独特的功能与优点。这些高档客房都拥有别致的设计和宽敞的布局，并可观赏内港和维多利亚其他别致建筑的壮观景色。每间宽敞、奢华的标准套房都采用开放的理念设计，每间客房都配备工作室、沙发床和别致的休息区，装饰十分优雅。一居室套房，宽敞明亮，每间套房都分为两个房间——一间卧室、一间客厅，中间用门隔开。客厅内配备一张沙发床，装修奢华，还有一些十分宽敞的客厅，客厅内摆放着滚动式折叠床。大部分套房都拥有一间浴室，一些套房内还额外配备了盥洗室。每间两居室套房都有三个房间：两间卧室和一个客厅。主卧室内摆放

着一张特大号床、次卧室内摆放着两张双人床，客厅配备了一张沙发床，此外，还留出一个十分宽敞的空间用来摆放滚动式折叠床。每间套房都配备一间私人浴室及全套的服务设施。

> Drawing inspiration from Queen Victoria's role as the Empress of India, this Bengal Lounge has colonial style.
> 以维多利亚女王被加冕为印度女皇时的时代背景为灵感，孟加拉休息大厅采用殖民地风格设计。

> Ground floor plan
> 一层平面图

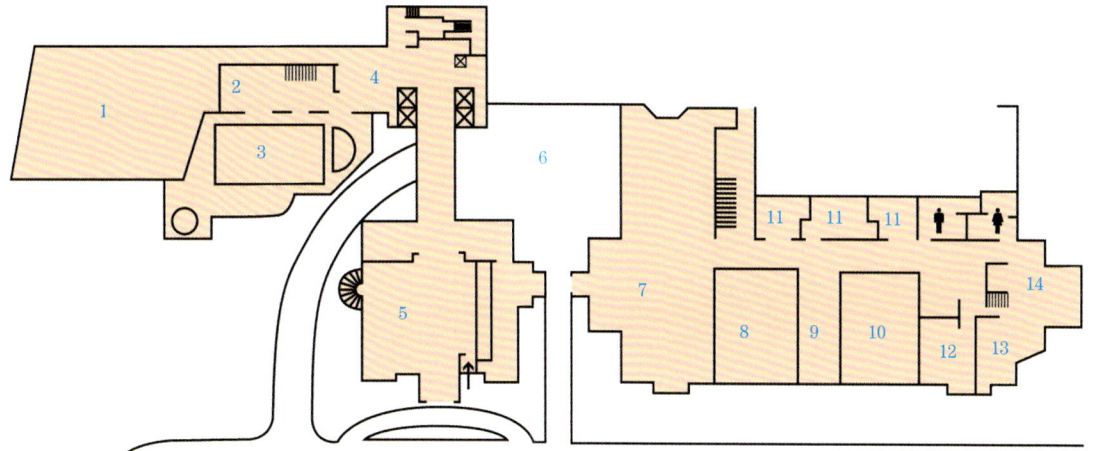

1. Willow stream spa   1. 柳溪水疗中心
2. Fitness centre   2. 健身中心
3. Pool   3. 游泳池
4. Elevators   4. 电梯
5. Reception lobby   5. 接待大厅
6. Shade garden   6. 树荫花园
7. Kiplings Room   7. 吉普林餐厅
8. Balmoral Room   8. 拜尔马洛宴会厅
9. Vestibule   9. 门廊
10. Buckingham Room   10. 白金汉宫宴会厅
11. Retail   11. 零售店
12. Kens-ington Room   12. 肯－辛通宴会厅
13. Windsor Room   13. 温莎会议室
14. Empress archives   14. 皇后餐厅

> The ceiling of the Library Lounge is hand-painted.
> The Empress Lounge is surrounded by rich chintz fabrics, antiqued tapestries and rugs, elegant wing back chairs.
> The picturesque backdrop of the Inner Harbour provides the quintessential Victoria experience to all who are enjoying the graceful dinner.

> 图书馆休息室的天花板采用手工绘制的图案。
> 皇后餐厅采用华丽的摩擦轧光织物、古老的挂毯和地毯及优雅的高背飞翼扶手椅作装饰。
> 内港餐厅如诗如画的背景带着每位享受优雅晚宴的客人回到了维多利亚时代。

> The Royal Table is an added refinement to guest's fine dining experience.
> Vintage furnishings make the Empress Restaurant elegant and classical.

> 皇室餐桌为游客精致的用餐体验增添了一丝优雅的氛围。
> 古老的家具为皇后餐厅增添优雅和古典的气息。

> The ballroom offers a variety of exquisite Victoria wedding locations to suit every size and occasion.
> Buckingham Meeting Room with marble fireplace is cosy and beautiful.
> The afternoon tea lounge features tea served in dainty William Edwards, (formerly Booth and Royal Doulton), china and sterling silver.

> 这间宴会厅拥有各种各样的婚礼场地，适合举办各种规模、各种形式的结婚典礼。
> 伯明翰会议室采用大理石壁炉作装饰，美丽温馨。
> 下午茶餐厅采用华丽的威廉·爱德华兹（原英国皇室道尔顿陶器制造商）陶器、陶瓷制品和纯银制品做茶具。

> In Gold Lounge, guests can unwind and enjoy special privileges such as access to various newspapers and magazines.

> Elegant, spacious and luxurious, Fairmont Gold One-Bedroom Suite has the added touch of a decorative non-working fireplace.

> A bedroom and a parlor is separated by a dividing door. The parlor room is lavishly furnished with a sofa bed.

> 在黄金休息室中，游客可以彻底放松，并可以阅读各种报纸和杂志。

> 费尔蒙特一居室黄金套房宽敞明亮、气氛优雅、装修豪华，还额外采用了一个装饰性的壁炉，为空间增加了一丝别致的氛围。

> 一扇门将卧室与客厅分隔开来。客厅中装饰着豪华的沙发床。

> These premier guest rooms are chosen for their distinctive room layout and spacious sizes, views of the Inner Harbour and distinctive architectures.

> The adjoining bedroom has a queen-size bed and a separate full bathroom that offers privacy from the other areas of the suite.

> The Signature Rooms have been personally selected for their unique features and benefits.

> 这些高级客房拥有独特、宽敞的房间布局、别致的建筑结构，并可一览内港的美丽景色。

> 临近的卧室拥有一张大号的卧床和一个独立的卫生间，卫生间与套房中的其他区域相独立，十分隐蔽。

> 招牌客房是设计师精心打造的客房，拥有别致的装饰和便利的设施。

FAIRMONT LE CHÂTEAU FRONTENAC

费尔蒙特芳堤娜城堡酒店

Completion date (项目建成时间) : 2007 / Location (项目地点) : Québec, Canada
Designer (设计师) : Wilson Associates / Photographer (摄影师) : Fairmont Hotels
&Resorts / Area (室内面积) : 33,000 sqm

Standing high on a bluff overlooking the mighty St. Lawrence River, Fairmont Le Château Frontenac is not merely a hotel located in the heart of Old Québec - it is the heart of Old Québec.

Fairmont Le Château Frontenac is Canada's most beloved hotel. In any one of the beautifully furnished accommodations, including 618 guestrooms and suites, guests will feel an elegant touch of historic Europe. The newly renovated accommodations offer an unparalleled level of luxury and service. With elegantly styled chateau furnishings and regal decor, each guest is treated to a spectacular stay in Québec City.

Deluxe Old Québec rooms are perfect for couples or single travellers. Finished in an elegant decor, these rooms overlook Old Québec's historical architectures such as the surrounding parks and the European style buildings of Old Québec City. Deluxe River View rooms are elegantly decorated in the traditional chateau setting. The feature of having a guaranteed view of the majestic St. Lawrence River enhances the unforgettable experience in a memorable setting. In addition to suites, Signature rooms are the best rooms in the hotel. Decorated in a unique chateau style, they represent the ideal accommodation associated with the prestige and reputation of Québec's Fabled Chateau.

Spacious and comfortable studio rooms are classic among families and long-stay guests. With an elegant décor, they offer a sitting area as part of the bedroom and can accommodate up to five people.

Charming and comfortable, Junior Suites are popular with families and long-stay guests. They are composed of a bedroom with one bathroom and a living room containing a sofa bed, a small hallway and the glass French doors separate the two rooms. Junior Suites are ideal for guests that would like to entertain as this room category offers sitting space for approximately five people in the living room.

Large and beautifully decorated, these suites combine the luxury and refinement of their experience with a spectacular view of the St. Lawrence River. They are composed of a master bedroom connecting to a large parlor with a sofa, arm chairs and a bathroom. Ideal for entertaining, the parlor sits approximately ten people.

> The entrance of the meeting room is grand and spectacular.
> 会议室的入口大气、辉煌。

费尔蒙特芳堤娜城堡酒店坐落于宏伟的悬崖之上，俯视气势磅礴的圣劳伦斯河。费尔蒙特芳堤娜城堡酒店不仅是坐落于魁北克古城中心的一家酒店，更是魁北克的核心建筑。

费尔蒙特芳堤娜城堡酒店是加拿大最受欢迎的酒店。酒店共有618间客房和套房，每间套房都装修精美，游客可在此感受到古代欧洲的优雅氛围。新装修的客房更是奢华至极，配套服务设施无与伦比。每一间客房都采用优雅的城堡式家具装饰，游客可在此观赏到魁北克城的壮观景色。

豪华的魁北克古典客房是情侣和单个旅行者的理想住宿场所。这些客房采用优雅的装饰，每间客房都可一览附近公园和欧式建筑等魁北克古城的历史性建筑。豪华的河景房采用传统的城堡式背景装饰，装修精美。房间内可感受到圣劳伦斯河的磅礴气势。奢华的背景、独特的观景体验令游客流连忘返。除了诸多套房外，标志客房是费尔蒙特芳堤娜城堡酒店的顶级客房。招牌客房采用别致的城堡风格装饰，理想的住宿环境与传说中魁北克城堡的威望和威严形成了匹配。

宽敞舒适的公寓客房是家庭游客和长住游客的最佳住宿选择。公寓客房装饰优雅，客房提供一个休息区，休息区还可用作卧室的一部分，能容纳5人。

标准套房迷人舒适，非常受家庭游客和长住游客的欢迎。标准套房配备一间带浴室的卧室和一个客厅，客厅内摆放着一张沙发床，旁边是一个小型走廊，两个房间用巨大的玻璃门隔开。标准套房拥有广阔的休息区，约容纳5人，是那些崇尚娱乐生活人士的理想之选。

这些套房宽敞明亮、装修优雅，游客可在豪华精致的环境中欣赏圣劳伦斯河的壮观景色。套房内有一间主卧室，主卧室与一间宽敞的客厅相连，客厅内配备一个沙发、诸多扶手椅和一间浴室。客厅可容纳将近10人，是理想的娱乐场所。

> The lobby with wooden wall is simple and elegant.
> 木质墙壁包裹的大堂装饰简单、气氛优雅。

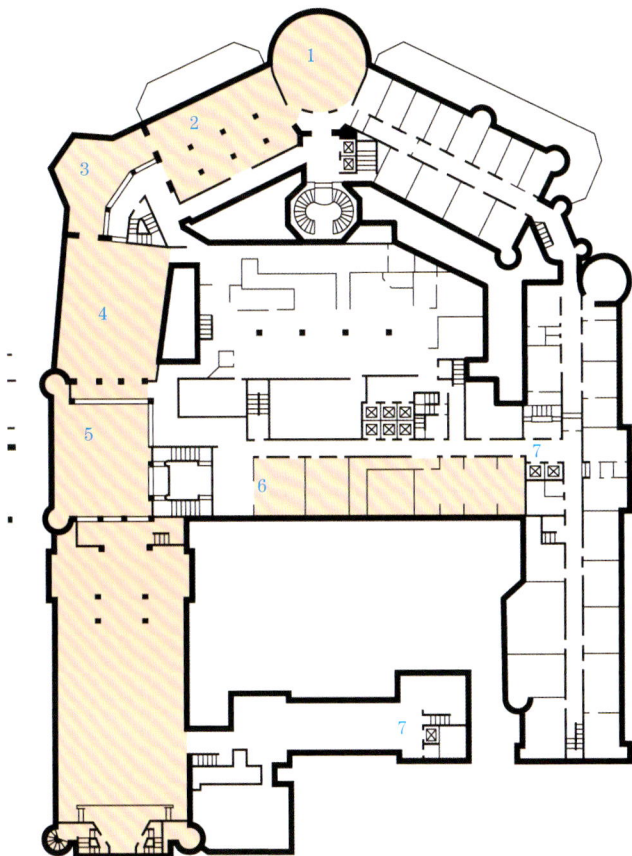

> First floor plan
> 二层平面图

1. Salon Rose   1. Salon Rose餐厅
2. Bellevue   2. Bellevue宴会厅
3. Petit Frontenac Salon   3. Petit Frontenac宴会厅
4. Frontenac Salon   4. Frontenac会议室
5. Vercheres Salon   5. Vercheres登记大厅
6. Montcalm Québec Salon   6. Montcalm quebec宴会厅
7. Elevators   7. 电梯

> Spacious lobby lounge has high windows, with the great view of the outside.

> An atmosphere of luxury is complemented by the glow of ten grand crystal chandeliers.

> The lavishly decorated castle ballroom is also an ideal place for meeting.

> 宽敞的休息大厅拥有高高的窗户，并将室外景色尽收眼底。

> 十盏华丽的水晶吊灯金光闪烁，烘托出更加浓厚的奢华氛围。

> 装饰奢华的城堡宴会厅是召开会议的理想场地。

> An atmosphere of luxury is complemented by the glow of ten grand crystal chandeliers, plush, elegant carpet, and a
  theatre stage draped with ivory velvet curtains.

> 十盏华丽的水晶吊灯、优雅的长丝绒地毯和采用乳白色丝绒窗帘装饰的戏剧舞台，烘托出更加浓厚的奢华氛围。

> With live piano music, the ballroom is also a convenient place for wedding.

> 宴会厅中响起生动的钢琴音乐，是举办婚礼的理想场地。

> This impressive architectural and rustic style room is ideal for big meetings, banquets, receptions and ceremonies.
> Fairmont Gold Lounge offers complimentary breakfast, all-day beverages and cocktail canapés served.
> Any occasion is a great occasion to enjoy the comfort and intimacy of Le St-Laurent Bar & Lounge with its sophisticated yet relaxed atmosphere.
> Amidst elegant and plush surroundings, Le Champlain is a stately dining room overlooking the St.Lawrence River.

> 这间乡村风格的会议室采用独特建筑结构设计，是召开会议，举办宴会、招待会和举行婚礼的理想场地。

> 费尔蒙特黄金大厅提供免费早餐、全天候的饮料和鸡尾酒沙发椅等服务与设施。

> 圣劳伦特酒吧与餐厅拥有优雅、轻松的氛围，您在任何时候都会沉浸在无比舒适和亲密的氛围中，体验完美时光。

> Le Champlain餐厅布满了毛绒装饰，气氛优雅，富丽堂皇，可一览圣劳伦斯河的美景。

> Le Champlain features wooden pillar and ceiling.
> The wooden fireplace is exquisite with engraving.

> Le Champlain餐厅拥有木质的柱子和天花板。
> 木质壁炉上方采用雕刻图纹作装饰，十分精致。

> Verchères Room is generously provided with windows for natural lighting and offers a view on Old Québec.
> Le Café de la Terrasse has a relaxed setting with an incomparable view on the St.Lawrence River and Dufferin Terrace.
> The Bibliothèque's roundness and elegant decor give to the room a prestigious and luxurious castle atmosphere.

> Verchères宴会厅拥有多扇窗户，既保证了充足的自然光照，又可一览魁北克老城的美丽风光。

> Le Café de la Terrasse餐厅拥有轻松的环境，餐厅中可一览圣劳伦斯河与都伏林步道的绝美景色。

> 图书馆的圆形布局与优雅的装饰为房间注入了高雅、奢华的城堡气息。

> Fairmont Gold rooms have all of the amenities for a relaxing stay amid understated elegance.
> Deluxe River View room is elegantly decorated in the traditional chateau setting, having a guaranteed view of the majestic St. Lawrence River.
> The Junior Suite has a separated living room offering sitting space for approximately five people.
> Fairmont room is ideal for couples or single travellers.
> Van Horne Suite in opulent decor including a master bedroom and a living room.

> 费尔蒙特黄金套房设施配备齐全，装饰朴素优雅，游客可在此享受轻松的时光。
> 豪华的河景房采用传统的城堡背景装饰，气氛优雅，客房内可一览圣劳伦斯河的壮丽景色。
> 标准套房拥有一间独立的客厅，客厅的休息区可容纳将近5人。
> 费尔蒙特套房是情侣和单身游客的理想下榻之选。
> 范·霍恩套房装饰丰富，拥有一间主卧室和一间客厅。

FAIRMONT CHATEAU LAURIER

费尔蒙特劳里尔堡酒店

Completion date (项目建成时间) : 2010 / Location (项目地点) : Ottawa, Canada /
Designer (设计师) : KWC Architects / Photographer (摄影师) :
Fairmont Hotels&Resorts / Area (室内面积) : 32,000 sqm

Located next door to the Parliament Buildings in the heart of Canada's capital - the landmark Fairmont Chateau Laurier Hotel in Ottawa is a magnificent limestone edifice with turrets and masonry reminiscent of a French chateau.

The design of Fairmont Chateau Laurier combines the ancient elements and dark warm colours, which make guests feel the smell of old times. Reflecting the confidence, dignity and style of Ottawa, Fairmont Chateau Laurier stands as a testament to this dynamic-thriving city.

Fairmont Château Laurier's original Ballroom, dating from 1921, was built in keeping with the style of a castle. It is a Renaissance-style ballroom with an 18-foot ceiling from which hang three spectacular crystal chandeliers that define every event as fit for royalty.

The most sought-after room in the city, the stunning Laurier Room features a 16-foot ceiling with gorgeous chandeliers, wall sconces, and Roman columns around the perimeter of the room. The cream and burgundy decor provides richness and warmth, making this the perfect venue for corporate events. The crowning glory of the Laurier Room is its access to a beautiful outdoor terrace featuring unparalleled views of the Ottawa River, Rideau Canal, and the Parliament Buildings.

Junior Suites offer a private living room separated from the bedroom by beautiful French doors. Corner Suites are bright and spacious. These suites have views of beautiful downtown Ottawa and Parliament Hill and are perfect for any occasion. One–Bedroom Suites are located throughout the hotel and consist of a bedroom connected to a parlor. The parlor is a separate room with its own sitting area, desk, bathroom and entrance from the corridor. The Presidential Suites are most luxurious accommodation. These suites are similar to a large one–bedroom apartment. The bedroom area consists of a dressing room, bureau, armoire with colour television and a large ensuite bathroom with an oversized shower.

> The hotel lobby is medium in size, with marble floors, highly ceilings, elegantly in the lobby and wool carpets round out the hotel's regal beauty.

> 中等大小的酒店大堂采用优雅的大理石地板和高高的天花板作装饰，地板上的羊毛地毯展现了皇室风格装饰的优美。

费尔蒙特劳里尔堡酒店位于加拿大首都渥太华市的国会大厦旁，是一栋雄伟的石灰岩建筑，搭配角楼和砌石工艺，仿佛法式城堡一般。

费尔蒙特劳里尔堡酒店将古典元素与深沉的暖色调相结合，使客人感受到古典的风韵。费尔蒙特劳里尔堡酒店彰显了渥太华特有的都市风格、城市的庄严与发展力，是极富动感、欣欣向荣的城市形象的象征。

费尔蒙特劳里尔堡酒店独创的宴会厅，始建于1921年，仿照城堡的风格设计。这个文艺复兴风格的宴会厅，拥有18英尺高的天花板，天花板上悬挂着3盏壮观的水晶吊灯，吊灯的大气辉煌与每一个适应皇室风格背景的主题活动相匹配。

该城市最受欢迎的劳里尔客房拥有16英尺高的天花板，搭配华丽的水晶吊灯、壁灯和房间四周古罗马时期的栏杆，雪白色与酒红色的装饰营造了华美与温暖的气息，这里是公司社交活动的理想场所。劳里尔客房最辉煌的设计就在于它与优美的室外露台相连，露台上可观赏到渥太华河、里多运河和国会大楼的壮观景色。

高级套房内还有一间私人客厅，美丽的落地双扇玻璃门将私人客厅与卧室相隔开来。拐角套房宽敞明亮，可一览渥太华市中心的美丽景色，是举行任何活动的理想场所。遍布酒店的一居室客房由一个卧室和一个客厅组成，客厅与卧室相连。总统套房是最豪华的客房。这些套房仿佛是大型的一居室公寓。卧室配备一间更衣室、一张办公桌、一个带彩色电视的大型衣橱和一间与卧室相连的大型浴室，浴室内配备一个超大型淋浴器。

> The Renaissance-style ballroom, dating from 1921, was built in keeping with the style of a castle.

> 这间文艺复兴时期的宴会厅始建于1921年，与城堡的建筑风格保持一致。

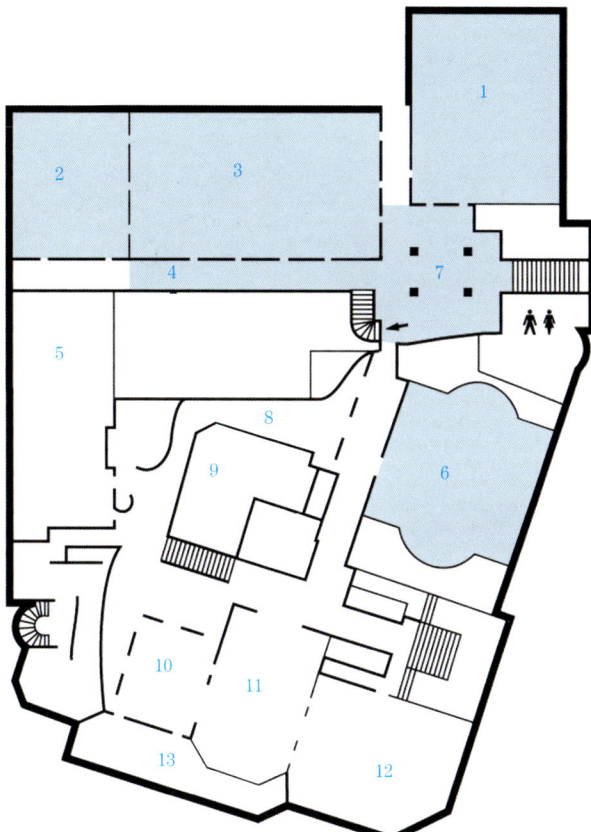

> Ground floor plan

> 一层平面图

1. Drawing room	1. 客厅
2. Laurier Room	2. 劳里尔客房
3. Ballroom	3. 宴会厅
4. French corridor	4. 法式走廊
5. Wilfrid's	5. 威尔弗里德
6. Adam Room	6. 亚当客房
7. Foyer of drawing room	7. 客厅的门厅
8. Peacock Alley	8. 孔雀厅
9. Retail	9. 零售店
10. Main lobby	10. 主大厅
11. Reading lounge	11. 阅览室
12. Zoe's Bar	12. 佐伊酒吧
13. Main entrance	13. 主入口

> Zoé's Lounge is now Ottawa's most stylish and elegant lounge with its chic decor, glowing chandeliers.
> High ceiling makes the grand ballroom solemn and luxurious.
> This ballroom has warm colour and soft light, and it is a perfect place for private party.

> 佐伊休息大厅拥有别致的装饰和金碧辉煌的枝形吊灯，现已成为渥太华最时尚、最优雅的休息大厅。

> 宏伟的宴会厅在高高的天花板的衬托下，显得更加庄严、奢华。

> 宴会厅拥有温暖的色调和柔和的灯光，是私人聚会的理想场地。

> In the exclusive Fairmont Gold Lounge, guests can enjoy coffee, tea in a tranquil atmosphere of relaxing comfort.
> Located within the elegant walls of Fairmont Château Laurier, Wilfrid's Restaurant offers a wide selection of culinary creations.
> The Zoe's Bar is closed to the window, and it's famous for its light fare menu.

> 在费尔蒙特黄金休息室中，您可在轻松、舒适、安静的氛围中品尝一杯咖啡和一杯茶。

> 被费尔蒙特劳里尔堡酒店优雅的墙壁包围的威尔弗里德餐厅提供各色经典美食，供您选择。

> 佐伊酒吧临窗而建，以其健康有机类菜肴著称。

> The living room of Presidential Suite that complete with faux fireplace, offers an exceptional environment to celebrate special occasions.
> The Fairmont Gold Suite living room has a French door.
> There is a reading lounge in main lobby, with the photo of Albert Einstein.

> 总统套房的客厅中采用了装饰性的壁炉，营造了别致的氛围，非常适合举行一些特殊的庆祝活动。

> 费尔蒙特黄金套房的客厅拥有落地双扇玻璃门。

> 在酒店的主大厅中有一个阅览室，里面还挂着阿尔伯特·爱因斯坦的照片。

> The Deluxe room is the essence of a Fairmont Hotels & Resorts experience, also featuring a cosy sitting area
  for rest and relaxation.
> Fairmont View Room is the most popular type of room, including an outstanding view of Canada's Parliament Buildings.
> The Fairmont Gold Room, with beautifully decoration, is equipped with a full marble bathroom.
> The newly renovated Executive Suite is separated by French doors to allow for additional privacy.

> 费尔蒙特酒店与度假村系列的每家酒店都以豪华客房为核心。豪华客房也有一个温馨、舒适、便于客人休息放松的客厅。

> 费尔蒙特眺望客房是该酒店最受欢迎的客房，客房内可一览加拿大议会大楼的壮观景色。

> 费尔蒙特黄金客房装修精美，客房中的浴室全部采用大理石为原材料。

> 刚刚翻新的行政客房与其他区域间用落地双扇玻璃门隔开，以便更好地保护客人隐私。

THE ST. REGIS
WASHINGTON, D.C.

华盛顿圣瑞吉酒店

Location (项目地点) : Washington, D.C, USA / Designer (设计师) : Sills Huniford
Photographer (摄影师) : The St.Regis Hotels&Resorts /
Area (室内面积) : 3,500 sqm

Located on K Street and just two blocks from The White House, this premiere DC luxury hotel is favoured by royalty, diplomats, and refined business travellers alike.

The hotel's luxurious Italianate exterior, dramatic public spaces, richly designed guest rooms and impeccable service create a luxurious, residential environment that have made this iconic hotel the destination for royalty, statesmen, business magnates, politicians and celebrities.

World-renowned interior designers Sills Huniford oversaw the redesign of the hotel's interiors, emphasizing elegant, modern luxury in custom furniture, fabric and technology. Known for their work of redesigning the acclaimed interiors at The St. Regis New York and with such clients as Tina Turner and Vera Wang, Sills Huniford re-imagined the refreshed guest rooms and suites in an incredible range of colour schemes. Steel blue, spring green, yellow daffodil, royal purple and rich rust are just a few of the colour palettes that guests will enjoy at the hotel. Custom-designed furniture mingles with original artwork and complements the deep history of the building.

Billowing lavender sheers bring warm light into the lobby during the day and transform the space with a deep glow in the evening and at night. The design team celebrated for their extraordinary sense of space, colour, texture and modernity, have created timeless spaces that embrace The St. Regis' legacy of unsurpassed luxury and trademark sophistication.

Each guest room and suite has a carved wood console built into the wall that serves not only as a beautiful design feature, but also holds the flat screen TV and disguises the closets, mini bar and safe. The ground level of The St. Regis Washington, D.C. is the home to a landmarked, hand-carved wood ceiling that was completely restored with new paint and gold leafing. Reminiscent of Italian palazzos, the ceiling is richly detailed and Sills Huniford masterfully chose fabrics and designs that would not compete with the ceiling but rather draws the eye upwards.

> The décor offers a compelling juxtaposition to the classical architecture, including the historic, landmarked ceiling.

> 包括具有历史意义的标志性的天花板在内，酒店的所有装饰赋予了这栋古典建筑以多重魅力。

> Lobby level plan

> 酒店大堂平面图

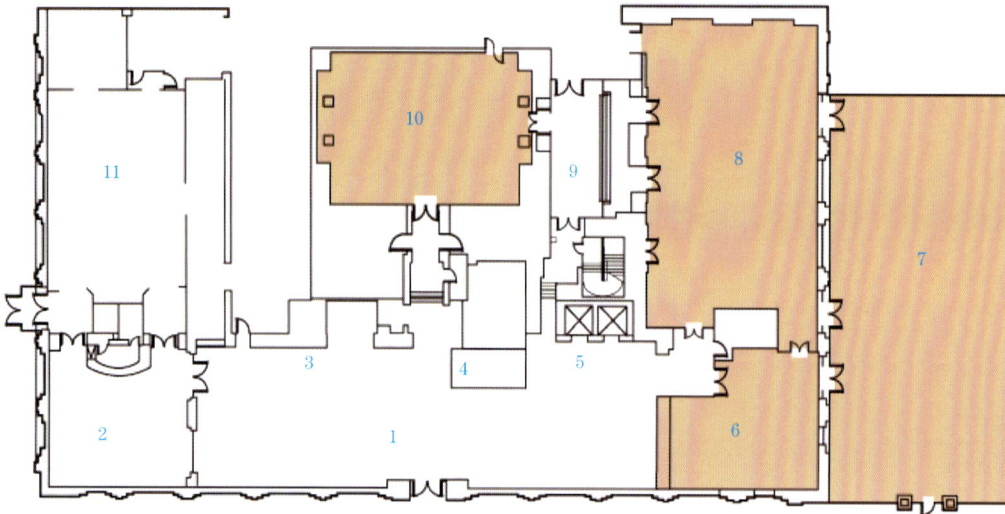

1. Lobby	1. 大堂
2. Lounge	2. 休息室
3. Concierge	3. 门房
4. Reception	4. 前台
5. Lifts	5. 电梯
6. Salon	6. 沙龙
7. Astor Terrace	7. 阿斯特露台
8. Astor Ballroom	8. 阿斯特宴会厅
9. Foyer	9. 门厅
10. Chandelier	10. 枝形吊灯
11. Restaurant	11. 餐厅

> The wall of Adour Restaurant is decorated with mirror.
> 阿杜尔餐厅的墙壁采用镜子作装饰。

华盛顿圣瑞吉酒店位于华盛顿的K大街上，与白宫仅隔两个街区，是美国首都最高档的酒店，用于接待皇室成员、外交官和品位高雅的商人等。酒店以意大利风格的奢华外观、设计精致的客房和卓越不凡的服务为游客打造了一个奢华的居住环境，从而使这家标志性的酒店成为皇室成员、政府首脑、商业巨匠、政界要员和社会名流的理想住宿之地。

世界著名室内设计事务所——思尔斯胡尼福德设计事务所负责指导监督整个酒店室内的翻新设计，采用了特制的家具、织物和技术，彰显出富贵高雅与现代奢华。思尔斯胡尼福德设计事务所因纽约圣瑞吉酒店的室内设计作品而闻名，这次与蒂娜·特纳和维拉·王等业主合作，采用了一系列大胆的色彩组合来装饰客房与套房。客人将在酒店中享受到色彩

的饕餮盛宴，其中包括钢青色、春黄色、水仙黄、贵族紫和铁锈红。特制的家具同原创艺术品搭配，与悠久的建筑历史相呼应。

如波浪般汹涌的淡紫色窗帘白天将温暖的光线带进休息室，傍晚和深夜，透过窗帘进入室内的昏暗光线又赋予了空间无穷的变化。设计师们对空间、色彩、纹理与现代性有很强的敏感性，从而打造了一个个永恒的空间，彰显了圣瑞吉酒店富饶的文化，体现出至尊奢华与品牌力量。

每一间客房与套房的墙壁上都安装了带有雕刻图纹的木柜，木柜不仅可以充当美丽的装饰，还可以承载纯平电视，或用作衣橱、迷你吧台或保险箱等。华盛顿瑞吉酒店的一层还搭配标志性的、手工雕刻的木质天花板，天花板外层采用新涂料和金箔装

饰，既防止腐蚀又美观大方。与意大利的宫殿造型类似，天花板拥有丰富的细节设计，而且思尔斯胡尼福德设计事务所特意挑选了一些织物与图案，它们既不会掩盖住天花板的风采又将人们的视线锁定在屋顶上方。

> The hand-painted ceiling in the newly restored lobby is very spacious.
> There are many precious artworks in the lounge.

> 新翻修的酒店大堂采用手工绘制的天花板，天花板十分宽敞。

> 酒店大堂中采用多种珍贵的手工艺品作装饰。

> One of entrance foyers of the Presidential Suite has graceful hand-painted wallpaper from China and other bespoke design details.
> The Empire Suite features a distinct green accented colour scheme, beautiful furnishings and decorative flourishes.
> The Caroline Astor Suite provides ample space for work or relaxation.

> 总统套房的一个入口门厅采用中国纯丝绸手绘壁纸和其他定做的细节元素作装饰。

> 帝王套房采用绿色作为主色调，精美的家具和装饰元素打造了大气辉煌的氛围。

> 凯若琳·阿斯特套房为游客提供了充足的工作与休闲空间。

CARLTON BAGLIONI HOTEL

卡尔顿巴格里奥尼酒店

Completion date (项目建成时间) : 2003 / Location (项目地点) :
Milano, Italy / Designer (设计师) : Reggiori / Photographer (摄影师) :
Baglioni Hotels / Area (室内面积) : 3,400 sqm

Situated in the vibrant heart of the city's shopping, business and cultural centre, the hotel faces right onto the most fashionable street of Milan.

The Carlton Hotel Baglioni is a truly city-type salon for those in search of a stylish and yet welcoming atmosphere; lavishly appointed with antique authentic furniture, marble mosaics and stylish drapes, the hotel lures guests with its sophisticated and elegant ambience. The Carlton Hotel Baglioni is also distinguished by prompt impeccable service.

Guest rooms at Carlton Hotel Baglioni are personalised with a passion for detail and elegant style and feature design that is both classic and contemporary.

The areas are spacious, inviting, perfectly sound-proofed and include the most up-to-date technology. The very luxurious suites of the hotel are decorated in original Art Deco Style. Each suite is personalised with unique pieces: handmade furniture, beautiful objects and warm-coloured silks.

The historical and well-known Il Baretto Restaurant is a meeting place for international politicians, journalists, fashion stylists and VIPs. Il Baretto serves light Milanese food and Mediterranean dishes, and enriches the range of services offered by the hotel.

The Caffé Baglioni offers a wide range of cocktails, appetisers and light food proposals that can be enjoyed in a warm and inviting atmosphere, or "al fresco" on the Baglioni Terrace. A well-known stopover for those people who working in Milan, the Caffé Baglioni is the ideal place for small meeting and top-level appointments, enhanced even further by the beautiful Library. The Library, stylishly furnished with beautiful pieces and elegant drapes, is welcoming at any time of day.

The Spiga 8 SPA at Carlton Hotel Baglioni is a urban oasis of psycho-physical wellness, where the guest can experience the unique Laura Elos method, based on the Mediterranean Thermal Cosmethic Line. Equipped with modern fitness equipment, showers, sauna, turkish bath and a lounge area for relaxing moments, the Spiga 8 SPA offers beauty treatments, holistic treatments, consultations with experts osteopaths and Day Spa.

> The reception is pleasant and welcoming, embellished with precious antiques and magnificent paintings.
> A renowned place of refreshment for those people who working in Milan, Caffè Baglioni, is ideal for small work meetings and major encounters.
> Caffè Baglioni with its large, light windows, provides a natural extension to the restaurant.

> 酒店前台采用珍稀古玩和辉煌的壁画作装饰，舒适宜人，给人一种宾至如归的感觉。
> 巴格里奥尼餐厅是放松身心的理想场地，受到米兰职场人士的热烈欢迎，非常适合举办一些小型的工作会议和招待活动。
> 巴格里奥尼餐厅拥有宽敞明亮的窗户，是餐厅的自然扩张。

卡尔顿巴格里奥尼酒店坐落于米兰市中心动感十足的购物、商业及文化中心，正对着米兰最时尚的商业街。

卡尔顿巴格里奥尼酒店是一个名副其实的都市沙龙，专门接待那些喜欢时尚与温馨氛围的宾客。酒店内采用经典的古典家具、大理石马赛克与样式新颖的窗帘作装饰，以其精致优雅的氛围吸引着八方游客。

该酒店拥有一间能容纳60人的大型会议室和4间能容纳8至25人不等的小型会议室。最具魅力的设计要属图书馆——采用英式风格设计，搭配古色古香的陈设，无论是休闲还是工作，这里都营造了独特的亲密氛围（小型会议室与大型的商务洽谈室）。

巴格里奥尼咖啡厅，采用柔和设计，是商务人士召开小型会议的理想之选，而历史上著名的Baretto al Baglioni餐厅则是米兰一家极富传奇色彩与历史意义的餐厅，这家餐厅原位于Via Sant' Andrea，现搬迁至著名的卡尔顿巴格里奥尼酒店。Baretto餐厅采用独特的风格装饰：古典的木壁、纯正优雅的英式家具、描绘打猎场景与骑马漫步英国乡村场景的挂画。

卡尔顿巴格里奥尼酒店的客房设计极具个性化，注重细节与特征设计，将古典与现代风格相融合。酒店内的32间古典风格的客房设计精致，设计风格主要通过细节和装饰得以体现。房间内配备意式家具、暖色调的真丝挂毯及与整个房间氛围完美融合的喷漆家具。宾客在露台上可俯视米兰最时尚的大街——Via della Spiga，无比优雅的意式风格布置在露台景色的映衬下显得极为别致。

> The Baretto can host up to 75 people, it feels like being in an English club house at the beginning of the twentieth century.

> Baretto酒吧能容纳75人，在这里用餐的游客仿佛置身于20世纪初的英国俱乐部。

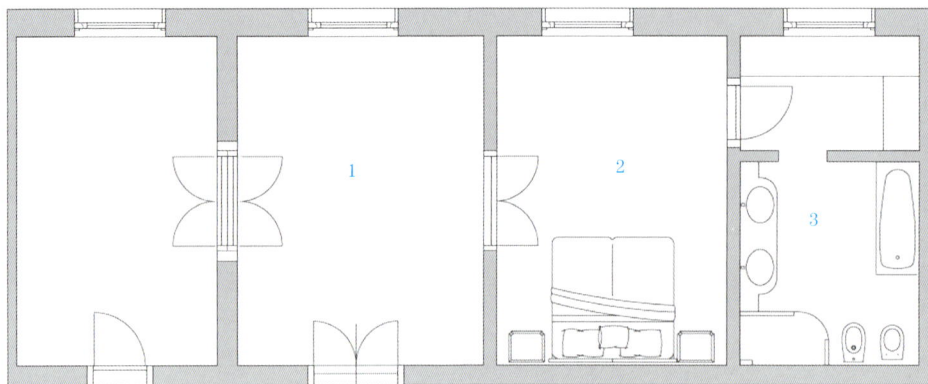

> Leonardo Suite plan

> 莱昂纳多套房平面图

1. Living room	1. 客厅
2. Bedroom	2. 卧室
3. Bathroom	3. 浴室

> This is the style of the Baretto: antique wooden friezes, elegant, typically British furnishing, paintings depicting hunting scenes.
> 这就是Baretto酒吧的风格：古色古香的木质雕带、优雅的特色英式家具及描述打猎场景的壁画。

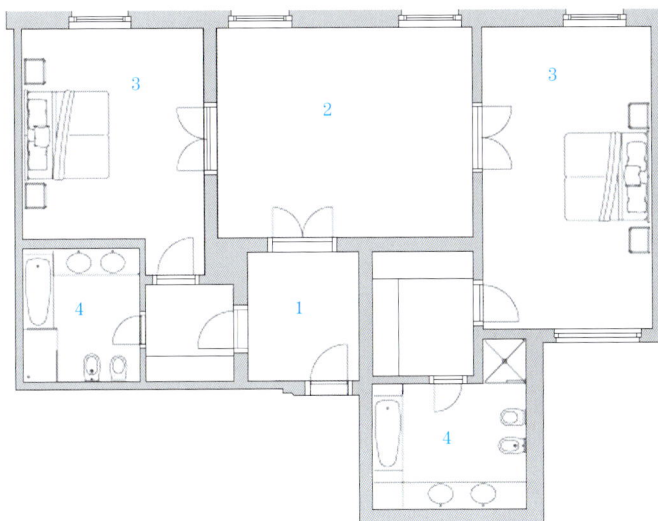

> Montenapoleone Suite plan
> 米兰之恋套房平面图

1. Entrance	1. 入口
2. Living area	2. 客厅
3. Bedroom	3. 卧室
4. Bathroom	4. 浴室

> The English-style library is complete with traditional furnishings.
> 英式风格的图书馆采用传统的家具装饰。

> The Carlton meeting room is warm and welcoming, with a passion for detail and elegant style.
> The modern convenience room also organising banquets and work dinner.

> 卡尔顿会议室装饰高雅，并注重每一处细节装饰，温暖舒适，令顾客感觉宾至如归。

> 配备现代化设施的多功能室也可举行宴会和其他工作聚餐。

> The most beautiful suite in the Carlton Hotel Baglioni has a welcoming day area with relaxing chairs and coffee table.
> The Superior Room is appointed with Italian furniture, silk tapestry, Murano hand-blown chandeliers.
> The Art Déco Suite combines different styles and culture to create a unique atmosphere, where Art Dèco style is mixed with ethnic details.

> 卡尔顿巴格里奥尼酒店最优美的套房拥有最赏心悦目的休息区，里面摆放着舒适的座椅和咖啡桌。

> 高级客房采用意大利家具、真丝挂毯和慕拉诺人工吹制的枝形吊灯作装饰。

> 装饰艺术风格套房融合了不同的设计风格与文化，营造了别致的氛围。在这里，装饰艺术与民族风韵完美结合。

> Art Déco Suite plan
> 装饰艺术风格套房平面图

1. Bedroom
2. Bathroom
3. Living room

1. 卧室
2. 浴室
3. 客厅

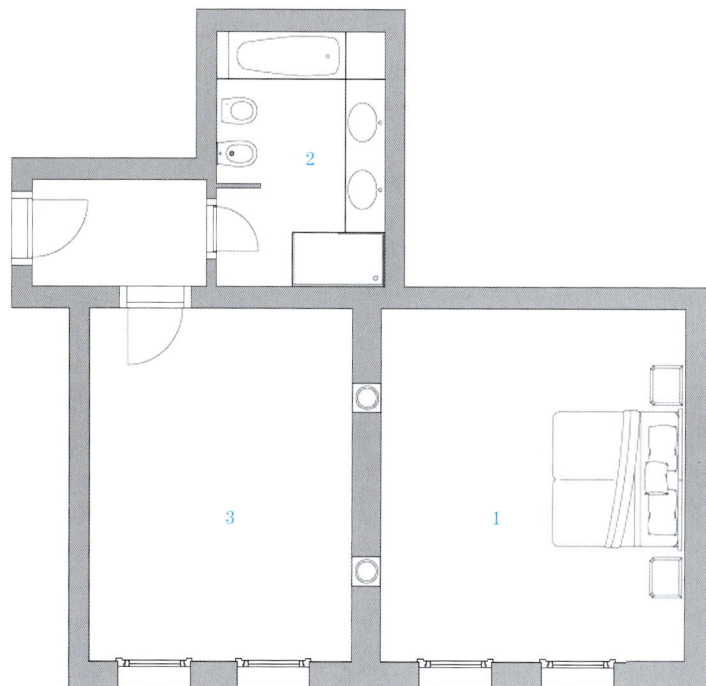

> The Deluxe Room is a combination of harmony and elegance thanks to the warm tones of the fabrics and beautiful furniture.
> The Classic Room of the Carlton Hotel Baglioni is marked by a discreet style with a great attention to detail and decor.
> The Superior Rooms are decorated with stuccos and silk tapestries and have carpet or parquet flooring.

> 豪华客房中织物的温暖色调与家具的精美打造了和谐、优雅的室内氛围。

> 卡尔顿巴格里奥尼酒店的经典客房非常注重细节与装饰，体现了无处不在的精致。

> 高级客房采用灰泥和丝绸挂毯作装饰，地面铺设地毯和镶花地板。

REGINA HOTEL BAGLIONI

巴利奥尼皇后酒店

Completion date（项目建成时间）: 2009 / Location（项目地点）: Rome, Italy /
Designer（设计师）: Carlo Busiri Vici / Photographer（摄影师）: Baglioni Hotels /
Area（室内面积）: 7,500 sqm

The Regina Hotel Baglioni in Rome is situated on the strategic, central Via Veneto, a stone's throw from the Villa Borghese, Piazza di Spagna and Via Condotti, the famous fashion street.

The Hotel Regina Baglioni represents a refined and unique combination, where luxury of the past continues in perfect harmony with a modern concept of cosmopolitan hospitality. The charm and simplicity of the façade combined with its refined interior furnishings, both in classic and Art Decò style, guarantees a unique atmosphere along with the reassuring feeling of being "home away from home".

As one enters the building, the elegant ochre columns and marvelous Art Nouveau staircase dominated by an artistic window come into view. The light from the grand Murano chandeliers, which hang from white stucco decorated ceilings, reflects on the marvelous geometrical figures of the marble floors from Siena. Some of his works of art are on display at the National Portrait Gallery of London, enhancing the historical and artistic inheritance of the hotel.

The elegance of the hall leads to the foyer, where the contrast between white stucco and violet walls along with the multifaceted patterned fabrics is elegantly combined with antique paintings and furnishings, making it an ideal place for special exclusive events.

The Brunello Lounge & Restaurant, with its exclusive independent entrance on Via Veneto, welcomes clients to a unique atmosphere, which unites traditional Italian elegance with contemporary design. It is the ideal location in which to sample dishes of Mediterranean flavour made from extremely high-quality products.

The Suites of the Regina Hotel Baglioni, some of which have private terraces with Jacuzzi, are facing the lively Via Veneto or overlooking the city. This accommodation offers guests a choice between a traditional Italian style or a more Art Dèco ambience. The Italian elegance is represented by antique furniture, silk tapestry and rich damask fabrics. Art Dèco interiors have sophisticatedly decorated marble floors, original furniture and details.

Here one can enjoy the serenity of a room, which is at the same time comfortable and sophisticated, aristocratic and friendly.

> The sumptuous marble lobby has classic columns.
> 奢华的大理石大堂拥有古典的栏杆。

> The Ludovisi Suite plan
> Ludovisi套房平面图

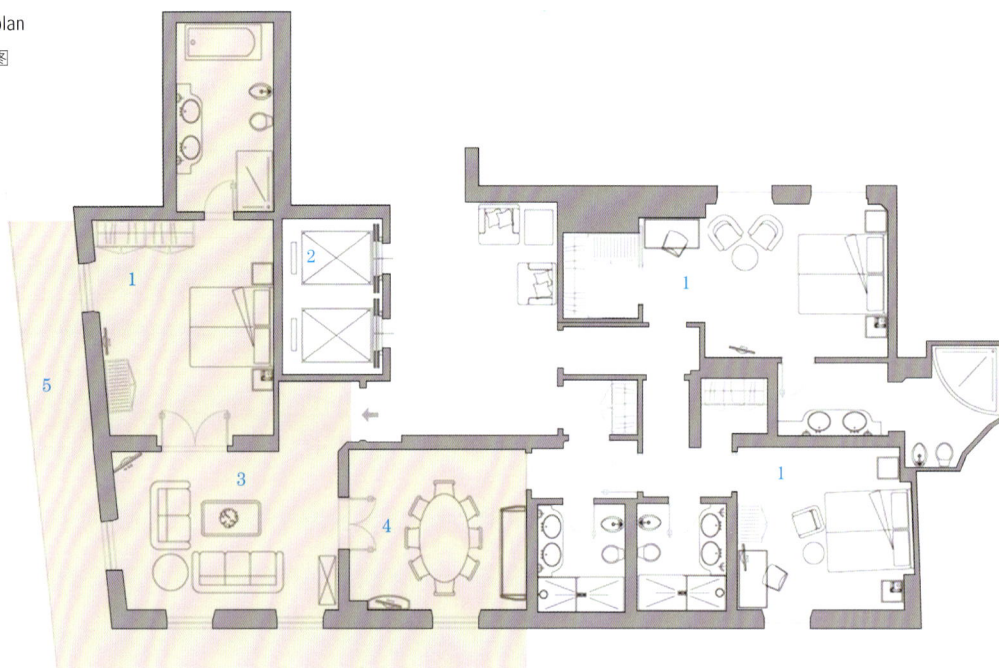

1. Bedroom	1. 卧室
2. Lift	2. 电梯
3. Living room	3. 客厅
4. Studio	4. 工作室
5. Terrace	5. 露台

> The Caffè Baglioni is the ideal venue for a distinguishing and luxurious gala dinner by an imposing working marble fireplace.
> 巴格里奥尼餐厅以熊熊燃烧的壁炉为特色，十分壮观，是举办别致奢华晚宴的理想场馆。

意大利罗马巴利奥尼皇后酒店位于罗马最主要的中心街道——威尼托大街上，距离鲍格才别墅、西班牙广场和著名的时尚街道——康多提大道仅一步之遥。巴利奥尼皇后酒店体现了古典奢华与现代都市酒店理念的完美结合，是一个优雅别致的组合体。建筑外观的简单与魅力，与精致的室内装饰相匹配。室内装饰采用古典与装饰艺术风格设计，别致的氛围令宾客身心放松，打造了宾客的第二个家。

游客一进入大厅，映入眼帘的就是优雅的赭色栏杆、新艺术风格的楼梯和散发着艺术气息的窗户。白灰泥装饰的天花板上悬挂着奢华的慕拉诺吊灯，灯光反射在锡耶纳大理石地板的几何图纹上。酒店的部分艺术品现在伦敦国家肖像画廊展出，这种文化遗产增加了酒店的历史意义和艺术气息。

装饰优雅的大厅直通门厅，门厅内的白灰泥墙壁与紫罗兰色墙壁形成了鲜明的对比，这种对比与多种图案的织物一起同古典的壁画与家具优雅的组合在一起，使这里成为举办重大活动的理想场所。

布鲁内罗酒吧与餐厅在维尼托大街有独立的入口，客人会在此体验到传统意大利式的优雅与现代设计的独特氛围。这里也是游客大饱口福的地方——游客将品尝到由优质材料制成的地中海风味特色美食。

巴利奥尼皇后酒店的部分套房带有摆放着按摩浴缸的私人阳台，阳台面朝维尼托大街，游客在上面可一览整个城市景色。酒店提供给游客两种住宿选择：传统的意大利风格的客房或装饰艺术气息更浓的客房。意大利式的优雅通过古典的家具、丝绸挂毯和质地丰富的锦缎得以展现，而装饰艺术风格的

客房则采用大理石地板、古典家具和细节设计等精心装饰而成。

在这里，人们可在舒适、经典、奢华和友好的氛围中享受房间的宁静。

> The small meeting room is decorated in brick dome.

> 小型会议室采用砖制圆屋顶作装饰。

> The Brunello Lounge & Restaurant unites traditional Italian elegance with contemporary design.
> 布鲁内罗餐厅将意大利的传统优雅与现代风格设计相结合。

> The Penthouse Panoramic Suite plan
> 屋顶全景套房平面图

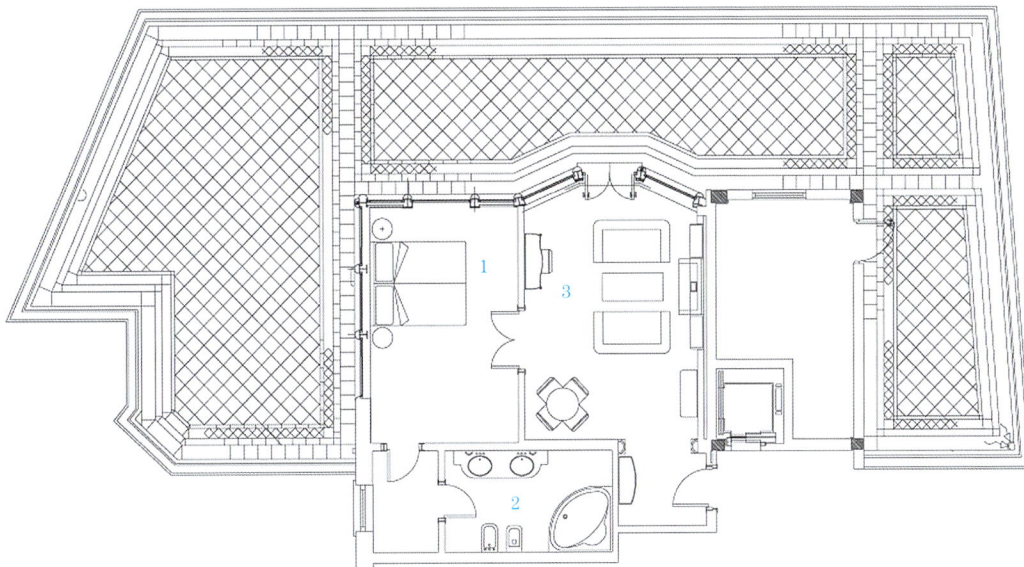

1. Bedroom   1. 卧室
2. Bathroom   2. 浴室
3. Living room   3. 客厅

> Standing on the stairs of the atrium overlooking the lobby, the guest can see marble floor reflection in the mirror.
> Caffe Baglioni that located in the hotel's elegant lobby area is warmed and enriched by the atmosphere of the cosy fireplace.
> Wellness area is an intimate location in which precious materials increase its sensory nature, and are warmed by light decorations.

> 站在中庭上方的楼梯俯视酒店大堂，客人能看到反射在镜子里的大理石地面。
> 巴格里奥尼餐厅坐落于酒店优雅的大堂区域，餐厅中舒适惬意的壁炉为整个空间增加了浓郁的温暖气息。
> 疗养区气氛温馨、亲密，采用珍贵材料设计，搭配灯光的装饰，增加了空间的感性。

> The Regina Suite plan
> 女王套房平面图

1. Bedroom
2. Living room
3. Bathroom

1. 卧室
2. 客厅
3. 浴室

> The characteristic of Penthouse Panoramic Suite is the Art Déco style with original furniture, Murano vases, and marble floors.
> The Art Decò style of Regina Suite is reflected through authentic briar , collector's items, and the distinctive black marble floors.
> Particularly fascinating light effects comes from the Swaroski's Cascade that dominates the living room.

> 阁楼的全景套房采用装饰艺术风格设计，以古典家具、慕拉诺花瓶和大理石地板为特色。

> 女王套房采用纯正的石楠木、精心收藏的古董和与众不同的黑色大理石地板作装饰，充分体现了装饰艺术风格的魅力。

> 客厅中的施华洛世奇水晶吊灯打造了整个房间中迷人梦幻的灯光效果。

> Two elegant columns separate the grace of the sitting room and working area from the intimacy of the bedroom.
> Ludovisi Suite is decorated in shades of dove grey and chocolate brown – the centerpiece is an impressive Murano glass chandelier.
> The Margherita Suite is located on the fifth floor of the Regina Hotel Baglioni offering a nice terrace.
> The comfortable living area boasts an ancient black marble fireplace with a classic-styled library for relaxing leisure time.

> 门将客厅、工作区与卧室分隔开来，客厅与工作区，气氛优雅，而卧室则拥有十分亲密的氛围。
> Ludovisi套房采用鸽子灰和巧克力棕色为主色调，中心装饰品是一顶华丽的慕拉诺玻璃水晶吊灯。
> 玛格丽特套房位于酒店的五层，拥有一个十分亮丽的阳台。
> 舒适的客厅拥有一个古老的黑色大理石壁炉，壁炉旁边是古典风格的藏书室，供游客享受悠闲时光。

THE LANDMARK LONDON

伦敦亮马酒店

Completion date（项目建成时间）: 2009 / Location（项目地点）: London, UK /
Designer（设计师）: HBA / Photographer（摄影师）: LHW /
Area（室内面积）: 18,000 sqm

The Landmark London is a five-star hotel on Marylebone Road on the northern side of central London, England, in the borough of London named the City of Westminster.

The hotel's design reflects the wealth and power of the era, and today's newly restored masterpiece will transport the guests back to a time of unimagined opulence and comfort. An outstanding feature of the hotel is the soaring glass-roofed atrium, providing a sense of space and light unique amongst London hotels.

The Landmark London's spectacular banqueting rooms are some of the finest in London. All rooms are classically designed with original Victorian features, very high ceilings, exquisite chandeliers and a vast amount of natural daylight. The hotel seamlessly integrates the romance and grandeur of the Victorian era.

Twotwentytwo Restaurant & Bar has been designed to ensure that when the guests visit, they will be enthralled by stunning surroundings. Twotwentytwo's design perfectly combines the traditional with eclectic modern features. The restaurant has original oak-wood panelling, sculpted ceilings, oversized mirrors and a wine library that showcases the hotel's large display of fine wines. The custom-made double-sided banquet seating provides a focal point, whilst bespoke pendants are suspended over tables featuring polished pewter tops with inset Champagne bowls. As a playful reference to the horse-drawn carriages that used to drop-off guests when the hotel was first built, they have a life-sized black horse lamp by Moooi at the hotel's entrance. Extra features include polished oversized mirrors and two specially commissioned art works next to original fireplace.

Designers have tried to keep elements of the hotel's distinctive Edwardian and art deco stylings while bringing everything up to date, and the owners have added an eight-room royal suite with an astronomical price tag of 10,000 pounds (nearly $16,000) per night.

With a five-red-star rating, the Landmark London ranks among the finest of the Capital's leading luxury hotels. With its own distinctive style and ambience, it combines classic British elegance and grandeur with the deluxe facilities demanded by today's discerning travellers.

> The music room includes an ornate sculptured ceiling and floor-to-ceiling windows, which ensure it is flooded with natural light.
> The elegant ballroom, with its classical chandeliers, is used to host a grand pre-dinner Champagne reception.
> It is a perfect venue with natural daylight, floor to celiling windows and magnificent chandeliers.

> 音乐厅中华丽的雕刻屋顶、巨大的落地窗使整个房间沐浴在自然光中。
> 优雅的宴会厅采用古典的枝形吊灯作装饰，将被用来举行晚宴前的香槟酒接待会。
> 自然的光线、巨大的落地窗和华丽的吊灯打造了一个完美的会议场馆。

伦敦亮马酒店是英国伦敦市中心北部的一家五星级酒店，酒店位于伦敦威斯敏斯特市的马里波恩路上。

酒店设计体现了时代的发展与繁荣，这个刚刚翻新的杰出作品将带游客体验无比奢华与舒适的时光。酒店最突出的特色就是带玻璃屋顶的中庭，屋顶高高耸起，形成一种强烈的空间感，而且照明条件在伦敦众多酒店中也堪称独一无二。

伦敦亮马酒店富丽堂皇的宴会厅是伦敦的顶级会餐场所之一。所有房间都采用古典设计，并带有原始的维多利亚特色——高高的屋顶、精致的吊灯和大量的自然采光。该酒店完美地将维多利亚时代的浪漫

与辉煌相结合。

Twotwentytwo餐厅与酒吧精心设计，保证住宿旅客一进门就被迷人的景色所吸引。Twotwentytwo餐厅与酒吧的设计完美地将传统特色与现代的折衷主义相结合。餐厅内装饰着古典的橡木嵌板、雕刻的天花板、超大号的镜子和一间白色的图书架，上面陈列着各色酒品。宴会厅两边定制的宴会椅是一大亮点，特制的垂饰悬挂于桌子上方，青灰色的光滑桌面带有内置的香槟酒杯。为与酒店成立之初接送客人的马车形成有趣的呼应，设计师还在入口处放置了由荷兰Moooi公司生产的、和原物一样大小的黑马落地灯。其他的特色布置还包括大型的抛光镜面和

古典壁炉旁两个特殊定制的工艺品。

设计师力求使用爱德华国王时代与装饰艺术主义的元素来装饰酒店，同时以最现代的手法加以展现。同时，业主又增加了一个八居室的皇室套房，套房的住宿价格十分高昂——每晚10,000英镑（相当于16,000美元）。

作为一家五星级酒店，伦敦亮马酒店是伦敦顶级的奢华酒店之一。酒店设计将古典的英式优雅与奢华同现代游客所追求的豪华设施相结合，打造了其独特的风格与氛围。

> Original oak panelling and traditional architectural materials make this drawing room a prefect venue in which to hold a ceremony.
> The Tower Suite accommodates up to 36 guests for lunch or dinner in a flexible choice of table configurations.
> The ballroom is an ideal venue for large gatherings, having the advantages of a high ceiling, plenty of floor space.

> 创新型的橡木嵌板与传统的建筑材料将这间会客厅打造成了一个举行典礼的完美场馆。

> 塔楼套房采用灵活多样的桌面布置，可容纳36名客人共进中餐或晚餐。

> 宴会厅中拥有高高的天花板和巨大的地面空间等便利条件，是举办大型聚会的理想场馆。

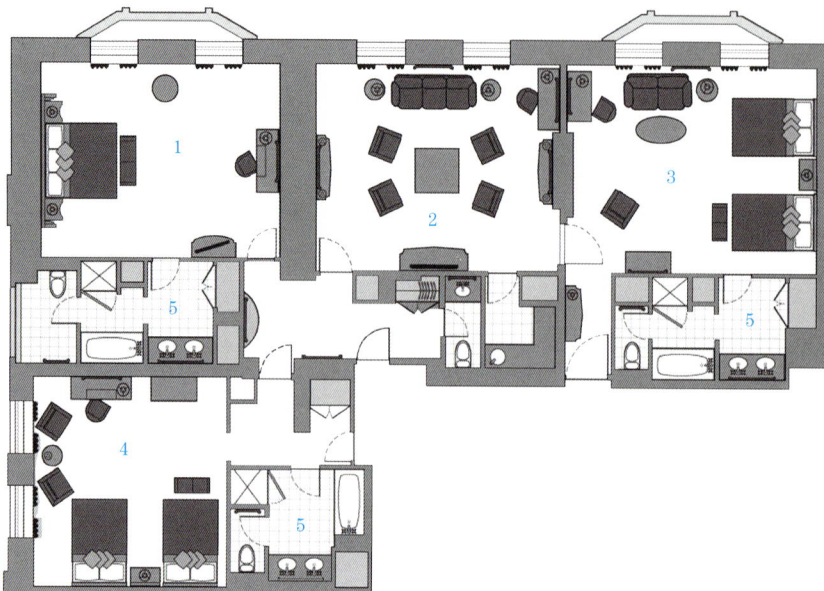

> Landmark Suite plan

> 招牌套房平面图

1. Master bedroom
2. Living room
3. Bedroom 2
4. Bedroom 3
5. Bathroom

1. 主卧室
2. 客厅
3. 卧室2
4. 卧室3
5. 浴室

> This Corner Executives suite boasts a king-size bed, generous storage space, a cosy sitting area with two-seat sofas and armchairs.
> The suite features a beautiful living area that is separated from the bedroom by a chic open archway.
> The newly refurbished Deluxe Suite is beautifully designed to create a relaxing and comfortable stay.
> This suite consists of a living area with a coffee table, a two-seat sofa and two armchairs.

> 景隅商务套房中拥有一张特大号床、宽敞的存储空间和舒适的客厅，客厅中摆放了双座沙发和扶手椅。
> 套房中优美的客厅与卧室间用一个别致的开放拱门相隔开来。
> 刚刚翻新的豪华套房设计精美，为游客打造了一个轻松舒适的下榻之地。
> 这间套房拥有一个客厅，客厅中摆放着一张咖啡桌、一个双座沙发和两把扶手椅。

HOTEL EXCELSIOR

怡东酒店

Completion date (项目建成时间) : 2009 / Location (项目地点) : Munich, Germany / Designer (设计师) : Jochen Dahms, Atelier Dahms, Tauberbischofsheim Photographer (摄影师) : LHW / Area (室内面积) :4,600 sqm

Situated next to Munich's main train station and a pedestrian walk to the Marienplatz, the Hotel Excelsior offers a convenient location in the heart of Munich.

Its interior design reflects old-world elegance, accentuated by brocade fabrics, antiquities and traditional Bavarian country style.

The Hotel Excelsior offers the Geisel's Vinothek, a famous restaurant offering Mediterranean delicacies and fine wines. The ornate lobby of the reopened Excelsior Bar is the perfect place for an evening drink.

This expansive suite of over 77 square metres offers spectacular views of the city from the top floor and invites the guest to experience the seamlessness of an ideally planned space.

The stunning bathroom, with luxurious fixtures and setting, celebrates privacy and opulence. The private terraces on both sides of the suite offer an even more unique perspective on this historic city.

It is a design hotel offering stylish rooms with a private terrace, flat-screen TV and free Wi-Fi. The air-conditioned rooms at the Hotel Excelsior feature minimalist design. They have tiled floors and modern furniture, and the private bathrooms come with a hairdryer. The 112 guestrooms including 8 Junior Suites, each truly remarkable, with soft lighting, elegant hardwood furnishings and gentle décor, provide ambient relaxation. Guests watch their worries drift away amidst the ultimate in quality, style and classic comfort.

Single Rooms provide the comfort and amenities that give a feeling of a second home. Both smoking and non-smoking rooms are available. Double Rooms combine a generous layout with well chosen amenities for a space that's both welcoming and quietly elegant. The Hotel offers both smoking and non-smoking double rooms that provide up to 25 square metres of space. Expansive suites combine space and luxury for a feeling of elevated elegance.

> Ground floor plan
> 一层平面图

1. Lift
2. Conference room 1
3. Conference room 2
4. Conference room 3
5. Lounge 1
6. Lounge 2
7. Men' Toilet
8. Women' Toilet

1. 电梯
2. 会议室1
3. 会议室2
4. 会议室3
5. 休息室1
6. 休息室2
7. 男士卫生间
8. 女士卫生间

怡东酒店坐落于慕尼黑市中心，交通便利，毗邻慕尼黑主火车站和一条通往玛利亚广场的人行道。
室内采用优雅的古典式设计，并采用锦缎、古董和传统巴伐利亚乡村风格进行点缀。
怡东酒店拥有著名的Geisel's Vinothek餐厅，餐厅提供地中海美食和上等的酒品。重新开放的怡东酒店内配备装修豪华的大厅，是晚宴的理想场所。
位于顶层的大型套房面积达77平方米，可一览整个城市的壮观景色。宾客可以在此感受精心设计的完美空间体验。

浴室内采用奢华的家具与布置，令人耳目一新，又具有极强的隐私性和富足感。
套房两边的私人阳台观赏到的城市风景更是别有一番情趣。
怡东酒店也是一个设计酒店，酒店带有一个私人阳台，配备纯平电视和免费的无线上网。酒店拥有一系列空调房，空调房采用极简派艺术风格装饰——瓷砖地板、现代家具，私人浴室内还配备吹风机。包括8间普通套房在内的112间客房都拥有柔和的光线、高雅的硬木家具和文雅的饰品，每间客房都

无与伦比，营造了轻松的氛围。游客在这个时尚、舒适、经典、高品质的天堂中会将烦恼抛到九霄云外，完全放松。
单人客房温馨舒适、设施配备齐全，仿佛是客人的第二个家。双人客房拥有大气的布局、精心挑选的配套设施，宁静优雅，给游客一种宾至如归的感觉。酒店客房分为吸烟客房和无烟客房两种，占据25平米的空间。宽敞的客房将空间与奢华相融合，营造了一种致尚的优雅。

HOTEL KOENIGSHOF

考宁蒴福酒店

Completion date（项目建成时间）：2009 / Location（项目地点）：München, Germany
Designer（设计师）：Jochen Dahms, Atelier Dahms, Tauberbischofsheim
Photographer（摄影师）：LHW / Area（室内面积）：4,500 sqm

Right in the centre of Munich, at the Karlsplatz/Stachus, the Königshof is located in the proximity of the most famous sights in Munich, including the best-known museums of the city.

The interior combines classic elegance with modern furniture design, including items from Donghia and selected Italian handcraft from Friuli and Lombardy.

In the hotel lobby, classic interior architecture and modern furniture designs combine to create an unmistakably amiable atmosphere.

The largest configuration of continuous salons offers 115 square metres of space. With natural light and a graceful ambiance, it's an ideal location for gathering.

With a variety of seating arrangements, this salon can welcome up to 110 people in a theatre-style setting or up to 100 people at tables. The salon can also be divided into individual rooms for separate workshop or presentation spaces.

In the suites, the traditional basic tenor of the furnishing is redefined with high quality textiles and pale colour-room compositions. Precious original pieces and the use of luxury materials represent the glamorous tradition of the house. The Suites define distinction and grace. The richness of the furnishings and amenities create an expansiveness that welcomes and delights.

These splendid accommodations offer a full and separate living room as well as visitor bathroom access. Configurations for two-bedroom suites are possible.

The Business Rooms pick up on the high demand placed on aesthetics and feel of the house and present themselves in a well-defined style of clear colours and lines. Rooms with extra-long beds–up to 2.1 metres. All rooms with halogen reading lights above the bed, individually regulated climate control, mini bar, safe, satellite television with pay TV, radio, wireless LAN with high-speed Internet access (free of charge), direct-dial telephones (also in some bathrooms), bathrooms with luxury shower or bathtub. There are also heated towel racks, cosmetic mirror and professional hair dryer.

The Deluxe Rooms are infused with a blend of elegance and sophisticated comfort. Each room invites the guest to relax into its generous, well designed space where taste and attention to detail blend to provide a sense of elevated luxury.

> The amazing restaurant has a great view of Munich.
> 迷人的餐厅中可观赏慕尼黑的美丽景色。

考宁葭福酒店位于慕尼黑市中心，毗邻卡尔斯广场与胡斯广场，周围是慕尼黑最著名的景点，其中包括慕尼黑最著名的博物馆。

室内设计将古典的优雅与现代的家具设计相结合，采用Donghia公司的家具与来自弗留利和伦巴第的意式手工艺品。

酒店大堂内，古典的室内设计与现代家具的完美结合打造了浓浓的亲密氛围。

一系列连续的沙龙，面积达115平方米，构成了酒店内最大型的结构。自然的灯光，优雅的氛围，是您下一次聚会的理想场所。

沙龙采用风格各异的席次布置，剧院风格的背景能容纳110人，或提供100人的座位。沙龙也可以被分为一个个独立的工作间或展示空间。

套房内，高品质的织物和白色的艺术品对家具的传统基调进行了重新界定。珍贵的古典饰品和奢华的材料体现了传统式住房的魅力。套房装修别致优雅。家具与设备的华美打造了温馨而喜悦的宽阔空间。此外，酒店还配备两居室的套房。

商务套房从美学和对房屋造型感觉的角度精心设计，清晰的线条与优雅的色彩打造了清新而经典的风格。套房内还配备了2.1米加长的床。所有房间的床头上方都配备卤素灯、个人温控设备、迷你酒吧、保险箱、带付费电视的卫星电视、无线电广播设备、高网速的无线局域网（免费）、直拨电话（一些浴室内也有直拨电话）及带奢华淋浴器和浴缸的浴室。此外，还有加热的毛巾架、化妆镜和专业的吹风机。

豪华客房将优雅、精致与舒适融为一体。每个房间都极具吸引力，令客人陶醉于大气、设计精致的空间氛围中，体验并关注每一处完美的细节，感受无比的奢华。

> The elegant meeting room brings graceful detail to private event.
> 装修精美的会议室突出了每一个优雅的细节，非常适合举办私人聚会。

> Ground floor plan
> 一层平面图

1. Entrance
2. Reception
3. Lobby
4. Lift
5. Bar

1. 入口
2. 前台
3. 酒店大堂
4. 电梯
5. 酒吧

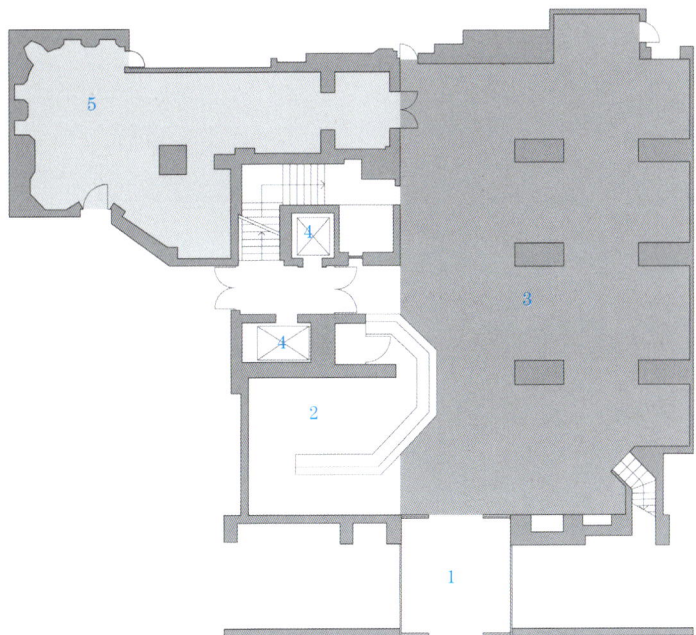

> The suite defines distinction and grace. The richness of the furnishings creates an expansive welcomes and delights.

> 套房的装饰体现了别致和优雅。豪华的家具营造了无处不在的欢乐友好氛围。

> The Deluxe Room is infused with a blend of elegance and sophisticated comfort.
> 豪华客房既拥有优雅的装饰，又拥有无比舒适的环境。

GRAND HOTEL MAJESTIC GIÀ BAGLIONI

吉亚巴利奥尼豪华大酒店

Completion date（项目建成时间）: 2007 / Location（项目地点）: Bologna, Italy /
Designer（设计师）: Alfonso Torreggiani /
Photographer（摄影师）: LHW / Area（室内面积）: 8,000 sqm

The Grand Hotel Majestic "Già Baglioni", the oldest and most prestigious hotel in Bologna, sits in the heart of the city on Via Indipendenza, within walking distance of Piazza Maggiore and the famous Due Torri.

The hotel has a total of seven meeting rooms, and four of which have been created recently. These rooms, all enjoying natural light, can each accomodate up to 120 people. The three original rooms are quite unique, each having the16th century coffered ceilings. The hotel is famous for its I Carracci Restaurant with its frescoes, which is rightly regarded as among the most sophisticated and elegant in Bologna. This is a place where people may enjoy the traditional recipes of the Emilia region whilst seated under authentic Italian masterpieces.

The 109 rooms have a mix of antiques furniture with modern amenities, a perfect blend of classic Italian style, elegance and hospitality. The fourth floor refurbishment has enriched the hotel with a new range of Junior Suites and Deluxe Room. The style in this case is classic the 18th century classical French.

All the rooms in the hotel are endowed with antiques and fitted with the latest modern conveniences. Many are in classical style, perfectly in line with the aristocratic feeling that the hotel emanates: wooden gems are woven together with beige and peach tones, creating a pleasant contrast with the contemporary style of the spacious bathrooms. They enjoy a magnificent view of the city from above, which takes in the roofs, monuments and beautiful historic centre of Bologna.

Superb drapery and paintings from the 18th century stretch down the hotel's long corridors and hang in the spacious rooms. The suites overlook the historic centre of Bologna. Beautiful draperies, silks and fabrics, paintings of the 18th century, finely polished marbles and precious mosaics, follow one another in the spacious hotel rooms and long tunnels.

Every suite is inspired by some of the most famous artists in the world, such as the celebrated musician Giuseppe Verdi who gave his name to the magnificent Royal Suite Verdi.

> Elegant carpet and warm light line its long corridors and spacious rooms.
> Europa meeting room is entirely dedicated to the wonderful frescoes of the
  school of the Carracci brothers.
> Vignola meeting room has high ceiling and window with elegant green curtain.

> 精致的地毯和温馨的壁灯装饰着长长的走廊和宽敞的客房。

> 欧罗巴会议室的天花板全部采用卡拉奇兄弟的神奇画作作装饰。

> 维尼奥拉会议室拥有高高的天花板和窗户，窗户上挂着优雅的绿色窗帘。

> The "Majestic Lounge-Café" overlooking original Winter Garden is characterised by original trompe l'oeil paintings.

> 宏伟的休闲餐厅俯视别致的冬季花园，餐厅中最突出的标志就是原创的错视派油画。

吉亚巴利奥尼豪华大酒店是博洛尼亚最古老、也是有名望的大酒店，坐落于市中心的独立路上，距离马焦雷广场和著名的迪特里酒店仅几步之遥。

该酒店共有7间会议室，其中4间为近期修建。这些会议室都拥有良好的自然采光，能容纳120人。三个原会议室装修别致，每一间都采用16世纪的方格天花板作装饰。

吉亚巴利奥尼豪华大酒店以其卡拉奇餐厅而闻名，餐厅内装饰着著名的壁画，这些壁画被认为是博洛尼亚技艺最精湛、最高雅的作品。在这里，你可以坐在纯正的意式壁画的下面，品尝艾米利亚地区的传统美食。

109间客房采用古典家具与现代设施的混搭装饰，打造了经典意大利风格、富贵典雅与好客氛围的完美组合。4楼经翻新后，增加了一系列普通套房和豪华套房。套房采用18世纪的古典风格和法式古典风格装饰。

酒店的所有房间都摆放着古董饰品，并配备最现代的便利设施。许多装饰都采用古典风格，与酒店散发的贵族气息完美匹配。木质珍品搭配米黄色与桃色基调，与宽敞浴室的现代风格形成绝美的对比。酒店顶楼可观赏包括建筑屋顶、纪念碑和博洛尼亚美丽的历史中心在内的众多壮观的城市景色。

18世纪华丽的织物与壁画装点着酒店长长的走廊。

套房内可一览博洛尼亚历史中心的美丽景色。绝美的布艺、丝绸和织物、18世纪的壁画、细致打磨的大理石与珍贵的镶嵌图案完美融合，共同装饰着宽敞的酒店客房。

每间套房的名字都来源于世界最著名的艺术家，比如辉煌的威尔第皇家套房就以著名作曲家威尔第的名字命名。

> The Art Deco Terrace Suite has a living room covered in precious Carrara marble in black and white tones with small silver points of light.

> 装饰艺术风格的露台套房拥有一间客厅，客厅地面采用珍贵的卡拉拉大理石，以黑白色调装饰，周围闪烁着银色的点点亮光。

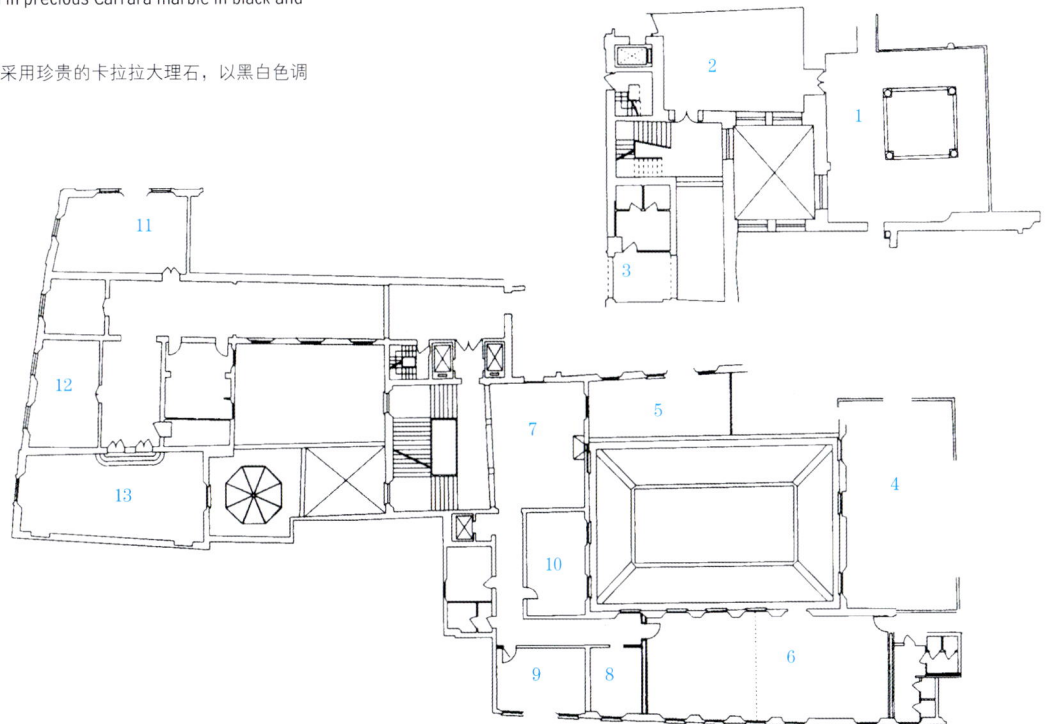

> Basement and first floor plan
> 地下室与二层平面图

1. 800 breakfast rooms	8. Nettuno	1. 800餐厅	8. Nettuno会议室
2. 900 breakfast rooms	9. Bentivoguo	2. 900餐厅	9. Bentivoguo会议室
3. Toilet	10. Pepoli meeting room	3. 卫生间	10. Pepoli会议室
4. Wellness area	11. Domenichino meeting room	4. 健身中心	11. Domenichino会议室
5. Terrace	12. Vignola meeting room	5. 阳台	12. Vignola会议室
6. Rinascimento	13. Europa meeting room	6. Rinascimento会议室	13. Europa会议室
7. Foyer		7. 门厅	

> The Delux room is lavishly furnished in a classic style, with an antique chest of drawers and a Murano-glass chandelier.
> The bedroom is enriched with soft chests in predominately black and silver-grey shades, and design elements.
> Large and very comfortable, the room is lavishly furnished in a classic style with a four-poster bed, an antique chest of drawers.
> New refurbished Junior Suite has French eighteenth century style.

> 豪华客房采用古典风格，装饰十分奢华，配备古典的衣柜和慕拉诺玻璃水晶吊灯。

> 卧室中布满了黑色与银灰色调的柔和衣柜与设计元素。

> 宽敞、舒适的客房采用古典风格的豪华装饰，卧室中摆放着一张四柱床和一个古典的五斗柜。

> 刚刚翻新过的标准套房采用法国18世纪风格。

HOTEL CIPRIANI

斯普莱利酒店

Completion date（项目建成时间）：2009 / Location（项目地点）：
Venice, Italy / Designer（设计师）：Michel Jouannet /
Photographer（摄影师）：Hotel Cipriani

Five minutes by private launch from San Marco, on the tip of Giudecca Island, this iconic Orient-Express hotel commands unrivalled views of the lagoon and Doge's Palace. Steeped in Venetian style, it is known for interiors decorated in exquisite local artefacts, classic cuisine with an innovative twist and the most fabulous swimming pool in the city.

One of Venice's most sumptuous dining rooms, the Fortuny is adorned with delicately blown glass, lustrous silks and amber-coloured mirrors. Magnificent views of the lagoon, stretching as far as the eye can see, and architectural touches reminiscent of St Mark's domes, combine to create the perfect setting for a relaxed candlelit dinner.

Hotel Cipriani offers 95 beautifully decorated suites and rooms, with magnificent views over the open lagoon and the gardens. Palazzo Vendramin is a 15th-century residence linked to the Hotel Cipriani through an ancient courtyard and a passageway lined with flowers.

Entering the Palladio Suite is like entering into another dimension: a glass capsule, suspended mid-air above the lagoon. One of the unique features of this Suite is a 180-degree view over the Venetian lagoon through floor to ceiling windows. The airy living room offers access to a private little balcony overlooking the lagoon. The elegant dining area allows guests to host small dinner parties served by a personal butler. Guests staying in the Palladio Suite have exclusive use of a large terrace with an outdoor heated long plunge pool and a heated Jacuzzi whirlpool, shaded by fragrant jasmine bushes. The master bedroom is sumptuously decorated and enjoys panoramic views across the magical lagoon.

The single rooms of Palazzo Vendramin, which are compact and charming rooms decorated in Venetian style, are ideal for the single travellers. Some can be connected to the suites, making them the perfect solution for families or large groups. All feature luxurious marble bathrooms.

The view from the Dogaressa Suite is akin to a Canaletto painting, the colours and enchanting details of St. Mark's Square framed by the imposing Doge's Palace and the vibrant Riva degli Schiavoni.

> Three chandeliers in the lounge make soft light.
> 休息室中的三盏枝形吊灯散发出柔和的光线。

斯普莱利酒店位于朱代卡岛顶端，从圣马可坐私人飞机仅5分钟的行程即可到达。在这家标志性的东方快捷酒店内可观赏到环礁湖与总督宫的绝美景色。酒店设计体现了纯粹的威尼斯风格，室内采用精致的当地手工制品作装饰，以创新型厨艺烹饪出的经典美食和威尼斯最美妙绝伦的游泳池为特色。

这里有威尼斯最豪华的餐厅——图尼餐厅，该餐厅采用精致的吹制玻璃、光亮的丝绸和琥珀色的镜子作装饰。一望无际的美丽礁湖景色与圣马可式的穹顶结构打造了完美的烛光餐厅背景。

斯普莱利酒店拥有95个装修精美的套房与客房，房间内可一览广阔的环礁湖与花园的壮美景色。文德拉明宫是一栋15世纪的住宅，一座古典的庭院和一条两旁长满鲜花的走廊将文德拉明宫与斯普莱利酒店相连。

走进帕兰朵套房仿佛进入了另一个维度：一个悬挂在环礁湖上方的空中玻璃体。这间套房的最大特色就是透过落地窗拥有观赏威尼斯礁湖的180度视角。客厅内空气畅通，可直通俯视环礁湖的私人小阳台。游客可在优雅的餐厅内举办小型的宴会，届时将有私人男仆为客人提供服务。帕兰朵套房内的客人可独享一个私人大阳台，阳台上还有一个长长的

室外热水游泳池和一个热水流的按摩浴缸，上方遮盖着芳香的茉莉花丛。主卧室装修豪华，可观赏整个环礁湖的神奇景色。

文德拉明宫内的单人客房紧凑精美，采用威尼斯风格设计，是单人旅行的理想住宿场所。一些客房与套房相连，非常适合家庭和大型的团体住宿。所有房间内都配备豪华的大理石浴室。

从公爵夫人套房内可观赏壮观的总督宫和包围在其中的圣马可广场的亮丽色彩与迷人的细节，还有斯其亚弗尼码头的繁华景象，仿佛是一副卡纳莱托画作。

> Very special and large Suite looks onto the open Lagoon and its islands.
> 在十分别致、宽敞的客房中可一览开放的环礁湖和周围各岛的美丽景色。

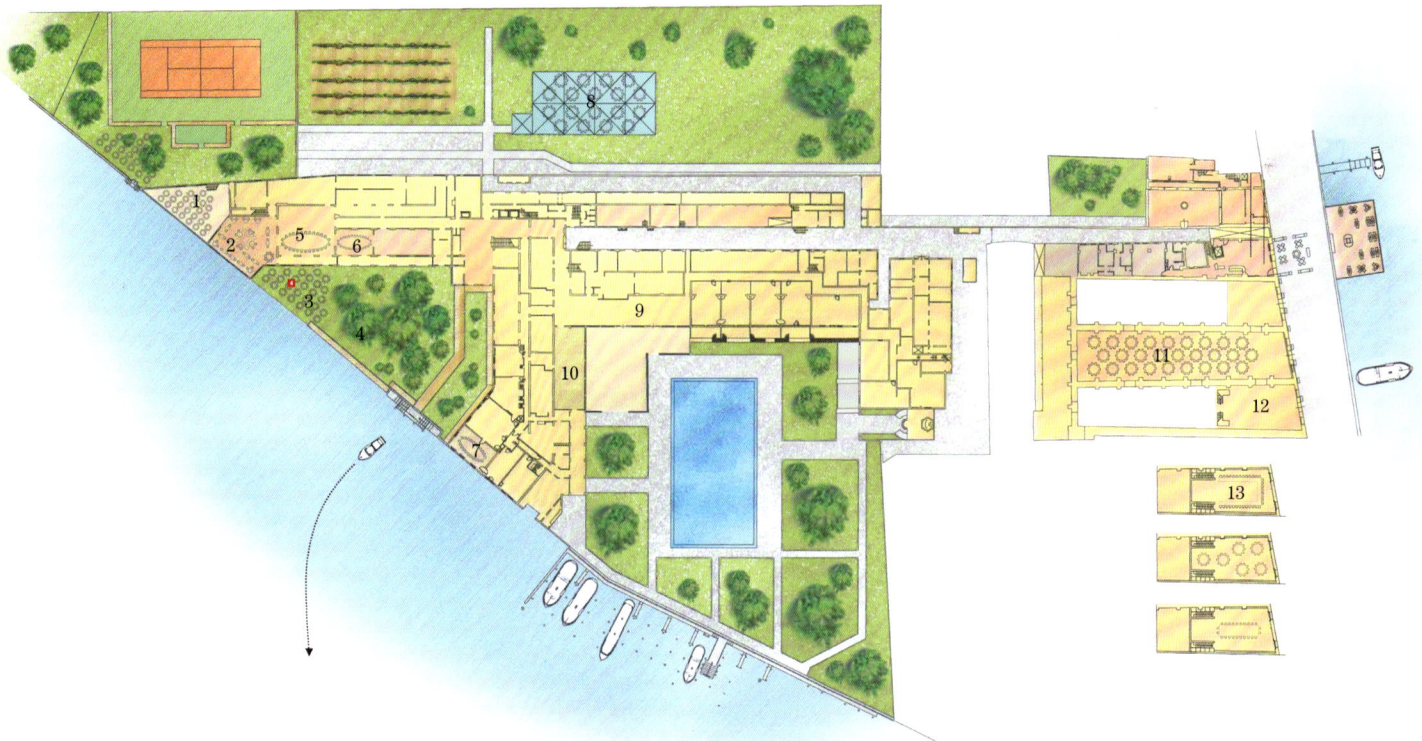

> Hotel siteplan
> 酒店总平面图

1. Antique garden	8. Casanova Gazebo	1. 古典花园	8. 卡萨瓦诺露台
2. Antique terrace	9. Pool loggia	2. 古典露台	9. 泳池边的凉亭
3. Fortuny Restaurant	10. Gabbiano Bar	3. 福尔图尼餐厅	10. 盖比亚诺酒吧
4. Fortuny Terrace	11. Longhi Ballroom	4. 福尔图尼露台	11. 隆基宴会厅
5. San Giorgio Room	12. Canaletto Room	5. 圣乔治奥宴会厅	12. Canaletto room宴会厅
6. Cigar Room	13. Canaletto Mezzanine Room	6. 吸烟室	13. Canaletto Mezzanine Room宴会厅
7. Guinness Room		7. 吉尼斯餐厅	

295

> Beautifully appointed, this Suite offers separate living room and provides all comforts.
> With a king-size bed, this peaceful room has precious painting.
> This compact and charming room, decorated in Venetian style, is ideal for the single traveller.
> The view from the Dogaressa Suite is akin to a Canaletto painting.

> 这间套房拥有独立的客厅，装修精美、无比舒适。

> 这间僻静的客房内摆放着一张特大号床，墙上还挂着珍贵的名画。

> 这间迷人的小型客房采用威尼斯风格装饰，是单身旅行者的理想下榻之选。

> 从Dogaressa套房中观赏的景色仿佛是卡纳莱托的一个杰出画作。

> The Suite marrying spaciousness with a sense of cosiness.
> Luxury Double room is decorated with a sumptuous crystal ceiling light.
>The Double room offers breathtaking views over the Csanova gardens.

> 宽敞舒适的套房

> 奢华的双人客房中挂着华丽的水晶吊灯。

> 双人客房中可欣赏到Csanova花园的迷人景色。

GRAND HOTEL EXCELSIOR VITTORIA

维多利亚怡东大酒店

Completion date (项目建成时间) : 2011 / Location (项目地点) : Sorrento, Italy
Designer (设计师) : Fiorentino family and various architects /
Photographer (摄影师) : Genius Loci / Area (室内面积) : 5,200 sqm

The 71 rooms and 26 suites are located in the three adjoining buildings of La Vittoria, La Favorita and La Rivale, which form the hotel complex.

The Excelsior Vittoria is ideal for conferences, business meetings, seminars or wedding receptions. The hotel's 6 elegant and well-equipped conference rooms combined with the staff's professionalism make a business gathering or celebration an enjoyable experience.

The elegance and luxury that characterise the Excelsior Vittoria is also reflected in the hotel's two restaurants, the Bosquet Terrace and the Orangerie poolside bar & restaurant. On the open-air Bosquet terrace with a spectacular view of Mount Vesuvius or in the park by the pool, guests can experience the highest traditions and quality of Neapolitan cuisine.

The most famous, the Caruso Suite, named after the famous tenor who occupied the room in 1921, boasts a large terrace with a sea-view and is kept in the very same style as it was when the famous Italian tenor lodged here in 1921 and there are endless stories about his stay like the piano, the writing board, some letters and photographs.

The Margaret Suite, one of the two "Antique" suites, takes its name from Princess Margaret who, on different occasions, stayed at the Excelsior Vittoria. The suite is decorated and furnished to preserve the charm of the early Victorian era. The living room has a painted neo renaissance ceiling and antique furniture from the early 19th century and the bathroom is in sumptuous local red marble with a beautiful cast iron enamelled bathtub.

The one unique and different from all the other rooms and suites of the Excelsior Vittoria is the new Designer Suite. The modern room, with four sunny balconies overlooking the Gulf of Naples and the hills of Sorrento, is conceived according to the contemporary Minimal Art style. The main element is the sun light, source of wellbeing, which radiates in the room thanks to the white walls, the hazel parquet floor, the varnished furniture and the wide glass-topped areas.

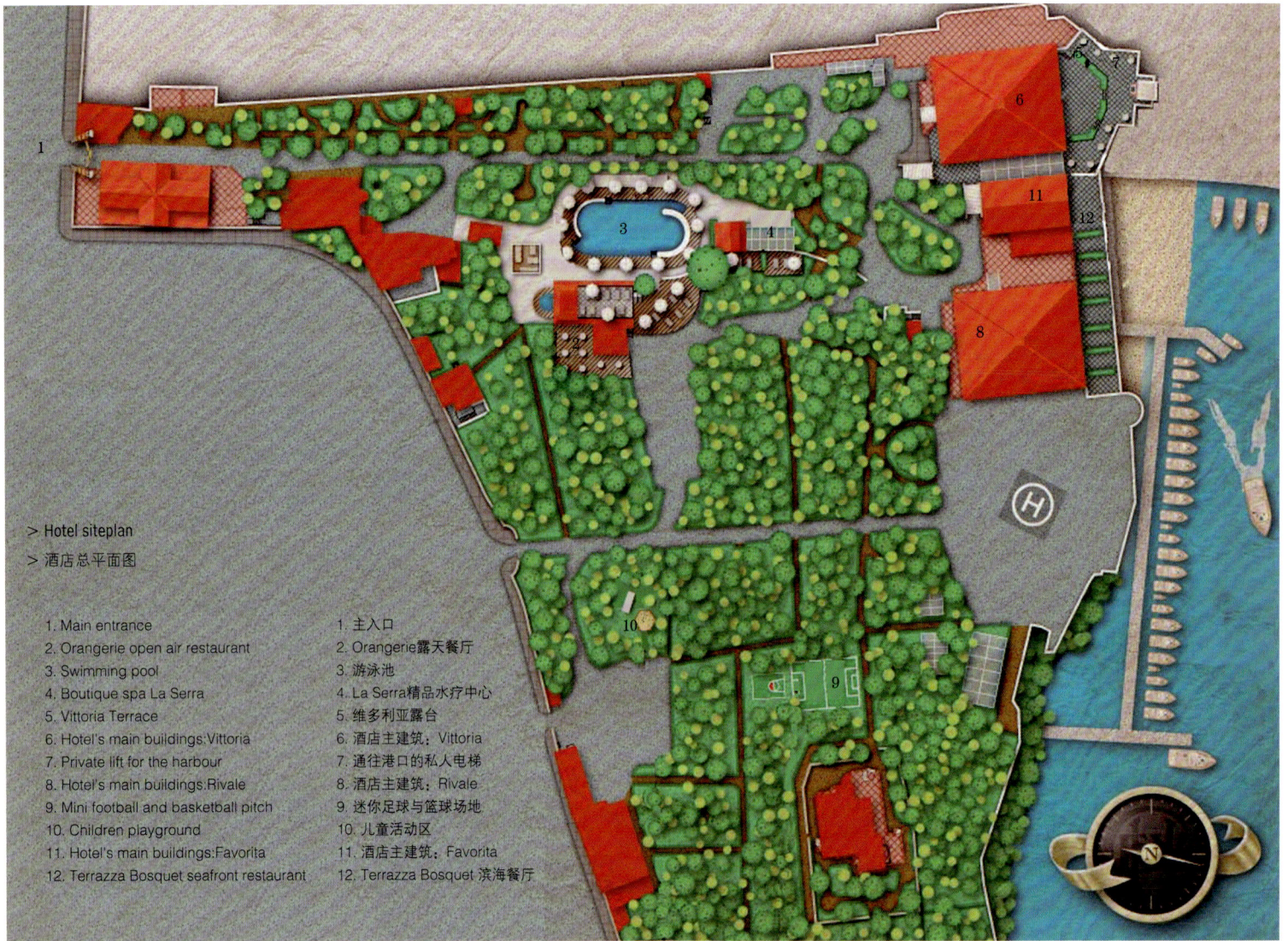

> Hotel siteplan
> 酒店总平面图

1. Main entrance	1. 主入口
2. Orangerie open air restaurant	2. Orangerie露天餐厅
3. Swimming pool	3. 游泳池
4. Boutique spa La Serra	4. La Serra精品水疗中心
5. Vittoria Terrace	5. 维多利亚露台
6. Hotel's main buildings:Vittoria	6. 酒店主建筑：Vittoria
7. Private lift for the harbour	7. 通往港口的私人电梯
8. Hotel's main buildings:Rivale	8. 酒店主建筑：Rivale
9. Mini football and basketball pitch	9. 迷你足球与篮球场地
10. Children playground	10. 儿童活动区
11. Hotel's main buildings:Favorita	11. 酒店主建筑：Favorita
12. Terrazza Bosquet seafront restaurant	12. Terrazza Bosquet 滨海餐厅

> The wholesome Mediterranean products and the magical backdrop of the Sorrento coast make Ristorante Vittoria one of the most beautiful delicious places for the bon vivant.
> The exquisitely frescoed ceilings and decorative furnishings create a sophisticated ambience exuding history.

> 健康的地中海特色美食和索伦托海滨的神奇背景使维多利亚餐厅成为最美丽的美食餐厅，专为那些追求高品质生活的时尚人士而打造。
> 精致的壁画屋顶和装饰家居打造了一个散发着古典韵味的雅致氛围。

维多利亚怡东大酒店是一个由三栋大厦——拉维多利亚大厦、拉法沃里达大厦和拉里瓦乐大厦组成的酒店综合体。酒店拥有75间客房和23间套房，这些客房与套房分别位于三栋大厦中。

维多利亚怡东大酒店也是召开会议、商务会谈和研讨会或者举办婚礼的理想场所。该酒店拥有6间装修典雅、设施配备齐全的会议室，所有工作人员都受过专业训练，从而使商务会谈或庆祝活动成为游客一次完美的体验。

维多利亚怡东大酒店的优雅与奢华从两个餐厅——丛林露台餐厅与橘园池畔酒吧及餐厅的装饰中得以体现。在开放的丛林露台餐厅中观赏苏威火山的壮观景色或在泳池边的公园中，游客既可品尝到高品质的那不勒斯美食，又可体验无比优雅的传统风韵。

该酒店最著名的套房——卡鲁索套房，以意大利著名男高音歌唱家卡鲁索的名字命名，他于1921年下榻该酒店。这间套房拥有一个大大的阳台，可观赏广阔的海景，并保持了卡鲁索入住时的装修风格，保留了如钢琴、写字板、一些信件和相片等小部件，这里无处不在讲述着当时的故事。

作为两个古老套房之一的玛格利特套房以玛格利特公主的名字命名。玛格利特公主当时会不时地下榻维多利亚怡东大酒店。这间套房保留了维多利亚早期的装修风格。客厅的天花板被涂成了新文艺复兴

时期的风格，家具则采用19世纪早期的古典家具，浴室采用当地奢华的红色大理石装饰，浴室内配备一个铸铁搪瓷浴缸。

在维多利亚怡东大酒店的众多套房与客房中，最具个性、最与众不同的要数刚刚修建的设计师套房。这间现代风格的套房带有4个阳台，阳台上可俯视那不勒斯湾和索伦托群山的壮观景象。套房以现代抽象艺术为理念构思而成。套房内最主要的装饰元素就是阳光。在白色墙壁、淡褐色镶木地板、涂漆家具和广阔的玻璃屋顶的作用下，阳光，作为幸福的源泉，遍布房间的每一个角落。

> The atmosphere in the suite inspired the song "Caruso", the famous tenor Lucio Dalla wrote in the late 1980s when he was a guest here.
> Suite Margaret is furnished with late-1800's furniture and has a terrace with sea view.
> With a sweeping terrace giving breathtaking views over the Gulf of Naples, the Royal Suite is furnished with period furniture, and French mirrors.
> Tastefully furnished, the fabrics and decor reflect the colours of the garden.

> 套房中的雅致氛围使人想起著名男高音歌唱家路西欧·达拉于20世界80年代末下榻该酒店时曾写的一首歌——"卡鲁索"。
> 玛格利特套房采用19世纪末的家具作装饰，并拥有一个海景阳台。
> 皇室套房拥有一个广阔的阳台，阳台上可一览那不勒斯湾的美丽景色，室内配备仿古家具和法式镜子。
> 室内采用雅致的装饰，织物和装饰都与花园的颜色相呼应。

> Overlooking the garden, this room is elaborately furnished and decorated with Renaissance style trompe l'oeils.
> Overlooking the Gulf of Naples, the classic room with sea view is furnished with fin de siècle pieces and elaborately decorated with geometric and floral stencils.
> Deluxe Premium, as the only sea-view room, has an open air private terrace elegantly furnished overlooking the bay of Sorrento and Gulf.

> 这间卧室俯视花园，装修精美，采用文艺复兴时期的艺术花纹作装饰。

> 高级豪华客房是唯一一间海景房，拥有一个优雅的露天阳台，阳台上可一览索伦托港与海湾的美丽景色。

> 古典客房俯视那不勒斯湾，可一览壮观的海景，室内点缀着19世纪末期的装饰元素，并采用几何图纹和花朵图案的蜡纸作装饰。

RELAIS SANTA CROCE

圣十字君主酒店

Completion date (项目建成时间) : 2009 / Location (项目地点) : Florence, Italy /
Designer (设计师) : Vasari / Photographer (摄影师) : Baglioni Hotels /
Area (室内面积) : 1,600 sqm

The palace in which the Relais of Santa Croce is situated was built in the early 1700s and was commissioned by the Marchese Baldinucci, powerful treasurer to the Pope and an important member of the Florentine aristocracy.

The imposing granite columns in the entrance hall were sent from Rome at his request and added to the austere and elegant appearance of the building.

Each floor has seven windows in the façade. Those on the raised ground floor are protected by austere railings and the stone frames extend down to the pavement and around the openings, which allow light and air to enter at basement level. The two entrance columns support a stone balcony above which hangs the coat of arms of the Balducci family, on which the inscription. In Deo Spes Mea can still be seen.

The suites at the Relais Santa Croce have up to 50 square metres of space and are composed of two separate rooms. The white lacquered wooden panelling sits well with the leather and cream and brown fabric of the armchairs and sofas, which provide a comfortable spot from which to watch a film on one of the large plasma screens. Every suite has an office space with desk and internet access.

The Relais Santa Croce has 24 rooms, and each one is different from the next. During the conversion process the primary objective was to create the greatest possible variety of accommodation in order to satisfy client demand and cater for individual taste; and at the same time make the most of the architectural value of the palace and its artistic riches and special atmosphere. Deluxe rooms, which have up to 30 square metres of space, are decorated in a variety of shades, from white to beige or green to orange. Some have an intimate feel with bleached parquet floors, others have charming specially-designed four-poster beds, each one furnished in a different style; the rooms on the top floor still have the original exposed roof beams, which allow for differing floor heights and romantic corners.

> The Musical Room is just wonderful to organise cocktails and exceptional dinners under the originals frescoes.

> 音乐厅采用古典的壁画作装饰，在高高的壁画下面举行鸡尾酒会和别致的晚宴，真是一种奇妙的体验。

圣十字君主酒店所在的宫殿修建于18世纪早期，宫殿主人是努奇侯爵——意大利教皇最有权势的财务长和弗洛伦萨贵族的一位重要成员。

按照主人的要求，酒店入口大厅壮观的花岗岩柱子从罗马引进，这些柱子增加了建筑外观的威严与优雅。

每一层立面都有7扇窗户。凸起的一层楼面上的窗户外部采用庄严的栏杆作保护，这种石头框架一直延伸至人行道与建筑的开口，这些开口可以使阳光照进地下室，并保证地下室内的良好通风。两个入口圆柱支撑起一个石质阳台，阳台上悬挂着巴尔杜奇家族的盾徽，盾徽上雕刻着"我信仰上帝"等字样的铭文。

圣十字君主酒店的套房面积可达50平方米，由两个独立的房间组成。白色喷漆的木制嵌板同扶手椅和沙发上的乳白色与棕色相间的织物完美搭配，客人可以躺在舒适的座椅上观看大型等离子显示屏上播放的电影。每间客房都有一个带办公桌和互联网接口的办公区。

圣十字君主酒店拥有24间风格迥异的客房。项目改建的最初目标就是要最大限度地增加客房的种类，以便满足客户不同的需求，迎合个人口味，同时实现对宫殿建筑价值、艺术财富和别致氛围的最佳利用。

豪华客房占地面积可达30平方米，采用白色、米黄色、绿色和橘色等不同窗帘装饰。一些客房内采用漂白的镶花地板，打造了一种亲密感；而另一些则采用特殊设计的四柱床，每一张床都以不同的风格装饰，十分迷人。顶层的房间内采用原始的外露屋顶梁，这些屋顶梁赋予建筑以变化的楼层高度和浪漫的角落空间。

311

> The breakfast room has natural light and high ceiling.
> 早餐室拥有自然的采光和高高的天花板。

> Each detail of Da Verrazzano Royal Suite was considered to create a unique atmosphere, where luxury and elegance harmoniously are united.
> Da Verrazzano 皇室套房的每一处细节都经过精心设计，营造了一种奢华、优雅的别致氛围。

> Da Verrazzano Royal Suite plan
> De Verrazzano 皇室套房平面图

1. Entrance
2. Living area
3. Bedroom
4. Bathroom

1. 入口
2. 客厅
3. 卧室
4. 浴室

> The rooms afford a view of the Santa Croce Basilica, the Cathedral dome or Florence's rooftops, and are decorated in different colours.
> The living room of Da Verrazzano Suite is very spacious, with exquisite ceiling.
>The room is characterised by great attention to detail: period furniture with design elements, precious fabric and luxury finishings.

> 客房中可欣赏到圣十字大教堂、教堂的圆屋顶或不同颜色的弗洛伦萨屋顶等美丽景色。

> Da Verrazzano 套房的客厅拥有精美的天花板，看上去十分宽敞。

> 客房设计突出细节装饰，采用带设计元素的仿古家具、珍贵的织物和奢华的饰物等作装饰。

> The Exclusive Room is decorated in various colours from white to red.
> This Deluxe Room has black wooden wall.
> The bathroom of Da Verrazzano Royal Suite has unique ceiling.

> 行政客房采用白色、红色等不同色彩装饰。
> 豪华客房拥有黑色的木质墙壁。
> Da Verrazzano 皇室套房的浴室拥有别致的屋顶。

SAN CLEMENTE PALACE HOTEL

圣克莱门特宫酒店

Completion date (项目建成时间): 2003 / Location (项目地点): Venice, Italy /
Designer (设计师): Carlo Busiri Vici
Photographer (摄影师): LHW / Area (室内面积): 14,000 sqm

The Island of San Clemente, a pearl of the Venetian lagoon, is situated in a peaceful but strategic spot, right in front of San Marco.

Two thirds of the Island's six hectars are green, openair spaces, which compose centuries-old park. Great care and respect have been kept to maintain the original typology of green areas: the perfumed linden trees – which have inspired "San Clemente Palace" fragrance by Laura Tonatto – cypresses, elms and nettle trees. Many are the courtyards among the Hotel's wings, such as the Clock Courtyard or the Plane Trees Garden.

"Gli Specchi" American Bar is located on the ground floor of the Hotel and provides a classical ambience, with the presence of many mirrors on the walls, which render the magical and sophisticated atmosphere. It is possible to taste refreshing cocktails, Venetian aperitifs, delicious Italian coffee and cappuccino, aromatic hot tea with pastries…

For guests loving the best in terms of comfort and width, the Classic Suite is the ideal category: it welcomes guests in a spacious entrance, it has high typical Venetian ceilings, a separated wardrobe area, an Italian marble bathroom, a sitting room separated from the bedroom by a sliding door.

The Executive Suite provides a warm entrance, a comfortable sitting room with sofa and armchairs, a living area with table to entertain friends and guests or to organise a small meeting, a bedroom reminding the Venetian style, and two bathrooms.

The Classic Double Room offers to its guests classical luxury and elegance, combined with sobriety of modern comfort. A romantic view of the centuries-old Park or of the internal gardens – the real symbol of our peaceful oasis – characterises the Classic Double Room. The Italian marble bathroom provides a shower area separated from the bathtub.

The Deluxe Double Room, with typical high Venetian ceilings, is characterised by a refined and elegant décor. Some Deluxe Double Rooms overlooks the Lagoon, others face the park, offering nice views of the gardens.

> Elegant public areas feeding off long corridors are adorned with fine antiques and luxury furnishings.
> The corridor in the lobby faces the lagoon with big tapestries.
> The lobby in medium size is elegantly furnished in a traditional style.

> 一条长长的走廊贯穿了优雅的公共区域，公共区采用珍贵的古董和奢华的家具作装饰。
> 大堂的走廊采用巨大的挂毯作装饰，大堂正对着外面的环礁湖。
> 中等大小的酒店大堂采用传统风格，装饰十分优雅。

圣克莱门特岛是威尼斯礁湖的一颗明珠，坐落于圣马可广场的正前方，宁静祥和而又占据具有战略意义的地理位置。

圣克莱门特岛占地6公顷，其中三分之二的面积为绿色的开放空间，还包括一个历史悠久的公园。威尼斯人对这个绿色空间给予了特殊的崇敬与关爱，从而保持了它的原始地貌：芳香四溢的椴树（以椴树的芬芳为灵感，罗娜·多纳多设计了圣克莱门特宫殿的芬芳气息）、柏树、榆树和荨麻树。许多绿地都被用作酒店翼楼的庭院，比如时钟庭院和悬铃树

园等。

坐落于酒店一层的美式酒吧Gli Specchi拥有古典的氛围，酒吧的墙壁上挂着很多面镜子，散发出神奇而又极富魅力的气息。游客可在此品尝到新鲜的鸡尾酒、威尼斯的开胃酒、香醇的意大利咖啡和卡普奇诺咖啡，还有茶香四溢的热茶与甜点。

商务套房拥有一个温馨的入口、一个摆放着沙发和扶手椅的客厅、一个带桌子的卧室，桌子可用来招待朋友、客人或在周围举行小型聚会，此外，还有一个威尼斯风格的卧室和两间浴室。

古典的双人套房呈现给顾客以古典的奢华、优雅与现代的舒适。双人套房的最大特色就是客人可以在客房中一览内部园林中的世纪花园——宁静绿洲的真正象征。意式的大理石浴室还有一个淋浴区，淋浴区与浴缸分离。

奢华的双人套房拥有高高的典型威尼斯式天花板，采用精致与优雅的装饰。一些奢华的客房可俯视整个礁湖的风景，另一些客房正对着公园，可观赏到花园的美丽景观。

> The elegance of Gli Arazzi dining halls is enhanced by authentic tapestries from the Royal Manufactory of Aubusson.
> The Oriente banquet room is decorated with original ancient Chinese furniture, and one of the walls is covered by ancient wooden doors.
> The Hotel features some small public sitting areas, which can be used for business appointments or reading area.
> The sitting area has comfortable sofa and chairs.

> 产自奥布松的纯正法式挂毯更增强了Gli Arazzi餐厅的优雅氛围。
> Oriente宴会厅采用中国古典的原创家具作装饰，其中一面墙壁上覆盖着古典的木门。
> 圣克莱门特宫酒店还拥有一些小型的公共客厅，这些区域可用来进行商务洽谈或用作阅读区。
> 公共客厅摆放着舒适的沙发和座椅。

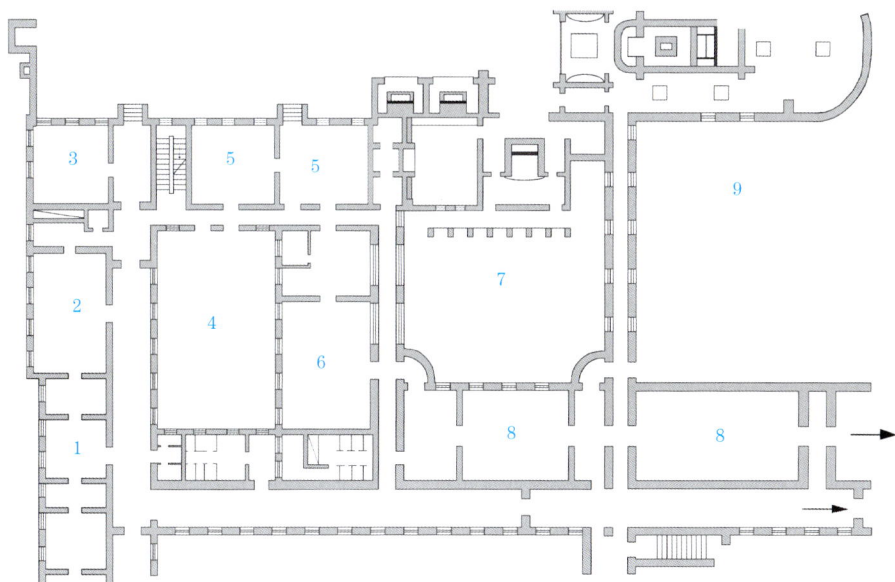

> Meeting room plan
> 会议室平面图

1. Lobby	1. 大堂
2. Gli Speechi Bar	2. Gli Speechi酒吧
3. Boutique	3. 精品店
4. San Benedetto	4. San Benedetto庭院
5. Le Maschere	5. Le Maschere餐厅
6. Oriente	6. Oriente会议室
7. The ancient cloister	7. 古典的回廊
8. Gli Arazzi	8. Gli arazzi美式餐厅
9. Della Monaca courtyard	9. Della Monaca庭院

> Classic Suite has a sitting room separated from the bedroom by a sliding door.
> The Deluxe Double Room, with typical high Venetian ceilings, is characterised by a refined and elegant décor.
> Deluxe Junior Suite has two windows, which gives a nice brightness to the ambience.

> 经典套房拥有一间客厅，客厅与卧室间用一扇滑动门隔开。

> 豪华双人客房拥有典型的、高高的威尼斯天花板，装饰精致、优雅。

> 豪华标准客房拥有两扇窗户，从而为房间氛围注入了亮丽的色彩。

THE FAIRMONT COPLEY PLAZA HOTEL

费尔蒙特科普利广场酒店

Completion date (项目建成时间) : 2010 / Location (项目地点) : Boston, United States / Designer (设计师) : Jinnie Kim Design / Photographer (摄影师) : Fairmont Hotels&Resort / sArea (室内面积) : 25,000 sqm

Centrally located in Boston's historic Back Bay, The Fairmont Copley Plaza sits steps away from the Boston Public Library, historic Beacon Hill, and the Freedom Trail.

Named "Best Steakhouse" by Boston Magazine, The Oak Room and the adjacent Oak Bar are two of the most elegant rooms in the city. Open for breakfast, lunch and dinner, the Oak Room is a regal, comfortably elegant restaurant that offers a traditional steakhouse menu with a distinctive twist. The Oak Room is adorned with gold, maroon and green fabrics and draperies that complement the dark wood panelling and mirrored walls. Cameos and lacunaria accent the intricate white plasterwork that details the 30-foot-high ceiling and twin Waterford crystal chandeliers illuminate the room.

The Oak Bar has been recently restored and is reminiscent of a British Officer's Club in the Orient. The same dark wood panelling and mirrors that flank the Oak Room are offset in the Oak Bar by smooth marble. Heraldry symbols decorate the ornately gilded and painted coffered ceiling. The furniture and décor create an inviting atmosphere to enjoy a relaxing drink or close a business deal.

The Fairmont Copley Plaza Hotel offers 383 individually decorated guest rooms, including 17 elegant suites.

Traditional guest rooms are elegantly decorated and inspired by the classic townhouses in Boston's Back Bay, featuring rich fabrics, custom-made furnishings and full marble bathrooms. All guest accommodations have been designed to reflect the grace of a glorious past, updated with contemporary conveniences.

Deluxe Room accommodations offer additional space and are individually appointed. Signature rooms are spacious junior suites designed to be the perfect retreat in the quiet inner courtyard of the Boston hotel.

The Deluxe One-Bedroom suites feature cathedral ceilings, decorative fireplaces, marble bathrooms, and a beautiful view of Boston's Copley Square. The spacious multi-room suites include a marble foyer, a luxurious bedroom and an elegant living room perfect for entertaining.

> The signature double-P crests decorate the mirrors and double doors, and the fish and shell motif is continued from the adjacent Oval Room.

> 镜子与双开门的上方装饰着标志性的双P顶饰，墙面同样采用了隔壁椭圆形会议室装饰用的鱼与贝克主题的图案。

> Main lobby level plan

> 主大堂平面图

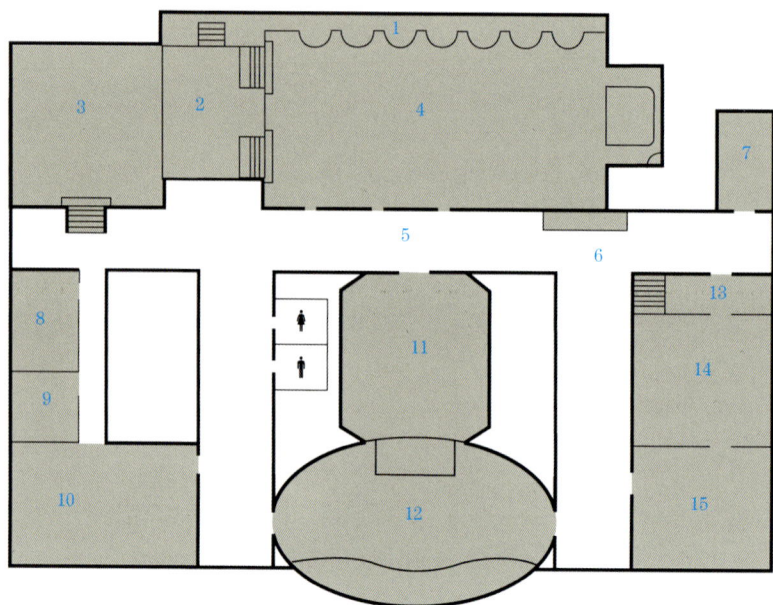

1. Balcony	1. 阳台
2. Rostrum	2. 讲台
3. Ballroom foyer	3. 宴会厅前厅
4. Grand Ballroom	4. 大宴会厅
5. Main lobby	5. 主大堂
6. Stairs to lower lobby	6. 通往楼下大堂的楼梯
7. Gift shop	7. 礼品店
8. Copley Room	8. 科普利宴会厅
9. Singleton Room	9. Singleton宴会厅
10. St.James Room	10. 圣詹姆斯会议室
11. Venetian Room	11. 威尼斯式婚礼宴会厅
12. Oval Room	12. 椭圆形婚礼宴会厅
13. Wine room	13. 酒室
14. Oak Room	14. 橡树屋餐厅
15. Oak Bar	15. 橡树酒吧

> The entrance hallway has been called Peacock Alley with its high dome and crystal chandeliers.
> 入口大厅拥有高高的圆屋顶和水晶吊灯，曾被称作孔雀厅。

费尔蒙特科普利广场酒店地处波斯顿历史上著名的后湾区，距离波斯顿公共图书馆、有着悠久历史的碧肯山和自由之路仅几步之遥。

该酒店的橡树屋餐厅和餐厅旁的橡树酒吧是波斯顿市两间设计最优雅的酒吧，曾被波斯顿杂志评为"最佳牛排餐厅"。橡树屋餐厅一日三餐向游客开放，采用王室的豪华装修风格，舒适优雅。这里既有传统的牛排美食，又提供独具特色的食品。橡树屋餐厅内，金色、栗色和绿色的织物与帷幔形成了对黑木嵌板和镜面墙壁的色彩补充。带浮雕的贝壳和各种格子图案装点着多种花纹交错的白色灰泥屋顶，天花板高30英尺，天花板上挂着两个沃特福德水晶吊灯，吊灯使整个房间熠熠生辉。

橡树酒吧最近刚刚经过整修，酒吧设计仿照了东方的英国官员俱乐部的风格。酒吧内也采用了装饰橡树屋餐厅两侧的黑木嵌板和镜子，同时又采用了光滑的大理石加以中和。镀金和涂漆的天花板上采用多个纹章作装饰。酒吧内的家具和装饰营造了一种迷人的氛围，游客可在此品尝一杯令人神清气爽的饮品或完成一笔商务交易。

费尔蒙特科普利广场酒店拥有383个独立装修的客房，其中包括17个优雅的套房。

传统客房装修高雅，设计灵感来自波斯顿后湾区的古典连排别墅——质地丰富的织物、特制的家具和大理石包装的浴室。所有客房都体现了古典的优雅，同时又提供现代化的便利设施。

豪华客房是专为个别游客设计的客房，客房内提供额外的独享空间。招牌客房都为宽敞的标准客房，这些客房被设计成了完美的休息寓所，坐落在波斯顿酒店幽静的庭院中。

奢华的一居室套房配备大教堂的天花板、装饰壁炉、大理石浴室，套房内可观赏波斯顿科普利广场的别致景色。宽敞的多室套房包括一个大理石门厅、一间豪华卧室和一间非常适合招待客人的优雅客厅。

> The elegant Grand Ballroom with a Louis XVI decor of crystal chandeliers and gilded decorations.
> The palatial Grand Ballroom is able to hold up to 1,100 delegates.
> Decorated in the rich classical style, which was popular at the turn of the century, the Oval Room is a more intimate room of symmetry and balance.

> 优雅的大宴会厅采用水晶吊灯和镀金饰品等路易十六时期的风格装饰。

> 宫殿般的大宴会厅能容纳1,100人同时用餐。

> 椭圆形会议室采用19世纪末20世纪初盛行的古典风格装饰，采用完全对称、平衡的结构，打造了更加亲密的室内氛围。

> The Venetian Room is Baroque-style architecture, filled with real and faux marble columns, matching chandeliers and wall sconces.

> 威尼斯宴会厅采用巴洛克风格的建筑结构，室内布满了人造大理石圆柱、与装饰风格相匹配的吊灯和壁灯。

> The Oval Room, considered one of the most beautiful rooms in Boston, features a realistic sky-and-cloud ceiling mural.

> 椭圆型宴会厅被认为是波斯顿最美丽的餐厅，以现实主义色彩的屋顶壁画为特色，上面画着天空与云朵。

> The Oak Bar is reminiscent of the World War II British Officer's Club in the Orient, with the same dark, rich panelling as the Oak Room and furniture upholstered in rich fabrics.

> 橡树酒吧仿照了第二次世界大战时期英国政府官员俱乐部的装饰风格，采用了与橡树屋餐厅同样的黑色豪华嵌板与家具，家具用豪华的织物作衬垫。

> The Oak Room is richly decorated with dark wood panelling and mirrored walls that are illuminated by twin Waterford crystal chandeliers.

> 橡树屋餐厅装饰奢华，采用黑木嵌板和镜壁作装饰，屋顶上方悬挂着一对沃特福德水晶吊灯，将整个空间打造得灯火辉煌。

> The Oak Bar features an extensive martini menu, as well as a fresh raw bar.
> The elegant surroundings of Wine Room, including a stained-glass window and dark wood panelling, are sure to impress any guest.
> The Oak Room provides luxurious upholstered banquettes that offer both comfort and privacy in a club-like atmosphere.

> 橡树酒吧以各式马提尼鸡尾酒为特色，还有一个生贝壳类食品自助柜。

> 藏酒室布置优雅，拥有一扇不锈钢玻璃窗和黑木嵌板，让所有游客耳目一新。

> 橡树屋餐厅拥有豪华的软垫座椅，打造了温馨而又隐蔽的俱乐部氛围。

> St. James Room offers a meeting space in a comfortable setting overlooking Copley Square.
> The oil painting on the wall creates classical atmosphere.
> The white marble fireplace creates cosy and classical atmosphere in Gold Lounge.

> 圣詹姆斯宴会厅拥有一间会议室，会议室温馨舒适，可一览科普利广场的美丽景色。

> 墙上的油画创造了古典的氛围。

> 白色大理石壁炉为黄金休息室打造了惬意的古典氛围。

> Fairmont Gold Lounge is an exclusive lounge modeled after a Back Bay Mansion.
> The spacious One-Bedroom Suite includes an elegant living room perfect for entertaining.
> The suite consists of a spacious living room with a decorative fireplace, including antiques and antique reproductions.
> A spacious marble entry foyer leads into an expansive living room of Presidential Suite with a decorative fireplace.

> 费尔蒙特黄金休息室是一个仿照后湾庄园住宅风格打造的专用休息室。
> 宽敞的一居室套房拥有一间优雅的客厅，这间客厅是完美的娱乐场地。
> 这间套房拥有一间宽敞的客厅，客厅中采用一个装饰壁炉、各色古玩和仿古玩作装饰。
> 宽敞的大理石门厅直通总统套房广阔的客厅，客厅中点缀着一个装饰壁炉。

> The guests can unwind in the gold library and enjoy special privileges such as access to various newspapers and magazines.

> 游客可以在黄金图书馆中身心得到彻底放松，还可享受到免费阅读报纸与杂志等服务。

> Fairmont Gold Library has elegant and noble cortex sofas.
> 费尔蒙特黄金图书馆拥有一张优雅、高贵的皮质沙发。

> Deluxe guest room features the same classic design as Fairmont rooms, inspired by Boston's Back Bay residences, while offering additional space.
> Each Fairmont Gold Room has been elegantly decorated in a rich classic design with custom-made furnishings.
> Each guest room has been elegantly decorated to be residential in design and inspired by the classic townhouses in Boston's Back Bay.

> 豪华套房与费尔蒙特套房采用相同的古典设计风格，仿照后湾庄园的住宅风格设计，同时又拥有额外空间。
> 仿照后湾庄园设计风格，每间客房都采用优雅的装饰，打造了住宅式的设计风格。
> 每间费尔蒙特黄金客房都采用特制家具，打造了奢华的古典设计风格，装饰十分优雅。

THE FAIRMONT HOTEL MACDONALD

费尔蒙特麦克唐纳酒店

Completion date（项目建成时间）: 2003 / Location（项目地点）: Edmonton, Canada
Designer（设计师）: Heather Jones & Associates / Photographer（摄影师）:
Fairmont Hotels&Resorts / Area（室内面积）: 13,000 sqm

Standing high on the bank overlooking the largest urban parkway in North America, The North Saskatchewan River Valley, The Fairmont Hotel Macdonald's charm and classic elegance have made it Edmonton's place for every occasion since 1915.

The hotel's dining establishments include the Harvest Room that offers panoramic views of the river valley and hotel gardens. The restaurant is the repository of museum quality art and antiques. The Confederation Lounge features leather furnishings and ornate wood panelling. Approximate 1,200 square metres of flexible function space accommodates gatherings. Health club amenities include an indoor saline pool, a spa tub, a steam room, a sauna and a fitness centre with cardiovascular and weight training equipments.

At the end of the day, there is no better place to relax than in the Confederation Lounge. This estate library-style lounge, with its plush chairs and welcoming atmosphere, is the best place to unwind with a beverage and admire the picture-perfect view of the largest urban parkway in North America - the North Saskatchewan River Valley.

This eight-storey hotel offers 198 guestrooms, all with two poster beds that feature triple sheeting, pillow top mattresses, 200 thread count linens and down duvets. Guestrooms boast three-metre ceilings and city views. Sophisticated decor includes armoires, upholstered armchairs and ottomans, and sunny colour schemes.

The Charles Melville Hays Suite is located on the eighth floor. This distinctive, elegantly appointed room features a comfortable sitting area and a double-Jacuzzi tub in the bathroom. The classic chateau ceiling follows the roofline of the hotel and has smaller windows with a view of the downtown city skyline.

Located on the eighth floor, the bright and open Lois Hole Suite is one of the most beautiful suites in the hotel. The bedroom features a king-sized bed with down-filled duvet and feather pillows. French doors open to the spacious living room, which has easy chairs, coffee and end tables, and double sofa bed. Both the bedroom and living room have televisions and large windows. A single-sized jetted bathtub is featured in the luxurious bathroom.

> The reception has high dome and beautiful carpet.

> 接待处拥有高高的圆屋顶和精美的地毯。

> Lobby level plan

> 酒店大堂平面图

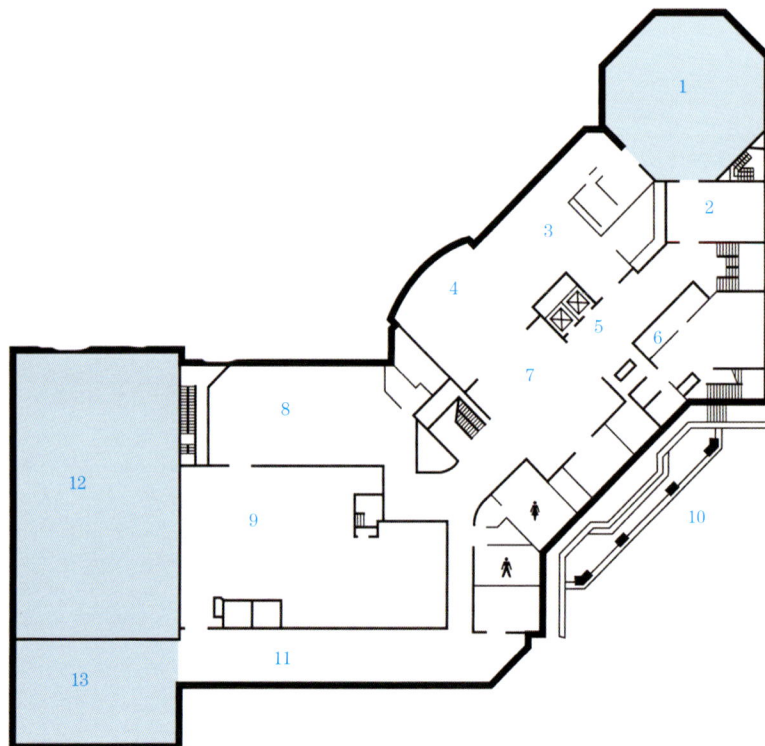

1. Wedgwood Room	1. 韦奇伍德宴会厅
2. Wedgwood Foyer	2. 韦奇伍德宴会厅门厅
3. The library lounge	3. 图书馆休息室
4. Confederation Lounge	4. 联邦大厅
5. Guest elevators	5. 客用电梯
6. Reception	6. 前台
7. Lobby	7. 大堂
8. Harvest Room	8. 丰收宴会厅
9. Service area	9. 服务区
10. Main entrance	10. 主入口
11. The galleria	11. 购物长廊
12. Empire Ballroom	12. 皇室宴会厅
13. Empire Ballroom foyer	13. 皇室宴会厅的前厅

费尔蒙特麦克唐纳酒店坐落在高高的河岸之上，俯视北美最大的城市林荫干道——北萨斯喀彻温堡河谷。自1915年以来，费尔蒙特麦克唐纳酒店就以其迷人的魅力和古典的优雅成为埃德蒙顿地区适合举办所有活动的高级酒店。

在费尔蒙特麦克唐纳酒店的诸多餐厅中，Harvest Room餐厅可一览河谷与酒店花园的全景。餐厅中陈列着大量博物馆级别的艺术品和古玩。联邦酒吧拥有皮革家具和华丽的木质护墙板。约1,200平方米的灵活功能空间可举办各种聚会。健身俱乐部中设施齐全，其中包括室内盐湖游泳池、温泉浴盆、蒸汽浴室和健身中心等，健身中心还配备心血管病的康复设备和举重训练设备。

日落时分，最好的放松地点就是联邦酒吧。这间图书馆风格的酒吧拥有豪华的长毛绒座椅，热情的氛围给顾客一种宾至如归的感觉，是游客舒展身心的理想场地。您可在此一边品尝饮料，一边欣赏北美最大的城市林荫干道——北萨斯喀彻温堡河谷如画般的秀美景色。

费尔蒙特麦克唐纳酒店共8层，拥有198间客房，所有客房都配备2张四柱床，床上铺着三层床单、高级床垫、经纬密度为200的亚麻织物及羽绒被。客房屋顶高达3米，可一览整个城市的全景。客房内部采用大型衣橱、加垫扶手椅、垫脚软凳等精致装饰和活泼的色彩设计。

查尔斯·梅尔维尔·海斯套房位于8层。这间优美高雅、装修别致的特制客房拥有一个舒适的休息区，浴室内还配备一个双人按摩浴缸。经典的城堡式屋顶装点着酒店屋顶的轮廓线，游客还可以透过小窗户看到整个市中心的城市轮廓。

同样位于8层的洛伊斯·霍乐客房是该酒店最美的客房。卧室内摆放着一张超大号床，床上铺着羽绒被，床头摆放着羽毛枕。客房通过法式对开门直通宽敞的客厅，客厅内摆放着安乐椅、咖啡桌、茶几和双人沙发床。卧室与客厅都配备电视和大型玻璃窗。奢华的浴室内突出摆放着尺寸均一的大理石浴缸。

> Richly carpeted marble lobby is adorned with glass chandeliers.
> The Wedgwood Room's round structure, domed ceiling, intricate Wedgwood detail, and gorgeous view of the
  North Saskatchewan River valley provide exquisite surroundings.
> The ballroom has breathtaking ceiling.

> 铺设华丽地毯的大理石大堂采用玻璃水晶吊灯作装饰。

> 韦奇伍德宴会厅拥有精致的环境——圆形的布局、拱形天花板、错综复杂的韦奇伍德陶瓷图案，并可一
  览北萨斯喀彻温堡河谷的壮观景色。

> 宴会厅拥有壮观的屋顶。

> The Harvest Room beckons with a warm and vibrant atmosphere, enhanced by a magnificent view of the North Saskatchewan River Valley.
> The Confederation Lounge is the best place to unwind with a beverage and admire the picture-perfect view of the North Saskatchewan River Valley.
> This estate library-style lounge with its plush chairs creates welcoming atmosphere.
> The Confederation Lounge with fireplace offers an elegance and warm atmosphere.

> 丰收宴会厅气氛温馨、活跃，宴会厅中可一览北萨斯喀彻温堡河谷的壮观景色。
> 联邦大厅是放松身心的最佳场地，游客可在此一边品尝饮料，一边欣赏北萨斯喀彻温堡河谷的壮观景色。
> 图书馆风格的大厅拥有长毛绒座椅，营造了热情友好的氛围。
> 联邦大厅采用壁炉作装饰，打造了优雅而温馨的氛围。

> The meeting room is richly decorated with oak panelling and woodwork.
> The suite has an elegant bedroom and a full bathroom.
> The Lois Hole Suite is one of the most beautiful suites. French doors open to the spacious
  living room, which has easy chairs, and end tables.

> 会议室采用橡木嵌板和木工制品，装饰十分奢华。

> 套房中有一间优雅的卧室和一个卫生间。

> 路易斯·霍尔套房是最漂亮的套房之一。落地双扇玻璃门直通宽敞的客厅，客厅中摆
  放着舒适的座椅和别致的茶几。

FAIRMONT ROYAL YORK

费尔蒙特皇家约克酒店

Completion date（项目建成时间）: 2003 / Location（项目地点）: Toronto, Canada
Designer（设计师）: Forrest Perkins / Photographer（摄影师）: Fairmont Hotels
&Resorts / Area（室内面积）: 70,000 sqm

Located in the heart of downtown Toronto, The Fairmont Royal York is within walking distance to the business and theatre districts, and the city's best shopping and dining area. With its grand stature and elegant interiors, The Fairmont Royal York depicts the elegance of its past while providing the finest in modern conveniences for today's travellers.

Within the past 10 years, over $125-million in renovations have restored the lobby, guest rooms and corridors, which included the addition of an award-winning Health Club, American Express Travel Service Centre, and the Executive Meeting Rooms on the 19th Floor. In 2001, a $12 million-dollar renovation resulted in establishing EPIC, a Four-Diamond

fine dining restaurant. This addition was followed by the completion of several other notable restoration projects: the Grand Lobby, the famed Imperial Room and its neighbouring lounge, The Library Bar.

Each of the 1,365 guest rooms at The Fairmont Royal York in Toronto offers accommodations that are luxuriously decorated and elegantly appointed. The Signature Rooms have been individually selected for inclusion in this premier guest room category on the basis of size, view and distinct characteristics. Capturing the essence of The Fairmont Royal York, each signature room offers elegant decor, spacious seating area and unique layout and design.

Executive Suites are popular with those wanting to hold a small reception.

Located from the eighth to the eleventh floor, they consist of two elegant bedrooms and a spacious living room with wet bar and conference table, which easily seats six people.

The Governor General Suite is located on the 16th floor. As guests arrive in this two-bedroom suite, they will be amazed by the majestic entrance foyer. The luxurious living room features a wet bar and decorative non-functional fireplace. Receptions of up to 16 people can be held in this suite. The hotel has one-bedroom suites with Small, Medium or Large parlors containing a sofa bed. The parlor is connected to a bedroom and separated by a single door.

> The lobby of the Royal York Hotel shimmers with light from large crystal chandeliers and is overlooked by the mezzanine level.

> 巨大的水晶吊灯使酒店大堂熠熠生辉，从夹楼可俯视大厅的全景。

费尔蒙特皇家约克酒店地处多伦多市中心，距离多伦多商业中心、剧场及多伦多最繁华的购物和餐饮中心仅几步之遥。费尔蒙特皇家约克酒店拥有宏伟的外观和高雅的室内装饰，再现了以往的优雅，同时又为现代游客配备了最便利的现代设施。

在过去的10年中，酒店的客房和走廊经过了大规模的整修，耗资1.25亿美元，其中包括曾获奖的健身俱乐部、美国运通旅行社和位于19层的常务会议室的补建工程。2001年，又斥资1200万美元，建立了一家四钻级高级餐厅。随后，又进行了几项著名的重建工程，其中包括大堂、著名的国王客房及客房旁边的酒吧——图书馆酒吧的翻新项目等。

费尔蒙特皇家约克酒店共1,365间客房，每间客房都采用优雅奢华的装修风格。招牌客房在房间大小、景色和特点方面都精心挑选，是首选的客房之一。每一间招牌客房都体现了费尔蒙特皇家约克酒店的精髓，拥有优雅的装饰、宽敞的座椅和别致的设计与布局。

行政套房最受那些渴望举行小型招待会的客人们的青睐。行政客房主要位于8层到11层，包括2间优雅的卧室和1间带小酒吧与会议桌的宽敞客厅，客厅能容纳6人。

总督套房位于16楼。游客一进入两居室套房，宏伟的入口门厅就会令其惊诧万分、赞不绝口。奢华的客厅拥有一个小酒吧和装饰性的壁炉。总督套房可容纳16人。

该酒店还提供大小不一的一居室套房，这些一居室套房还带有大、中、小等型号不一的客厅，客厅内摆放着一张沙发床。沙发与卧室相连，中间用一扇门隔开。

> Gleaming marble, polished wood paneling, rich draperies and the warm smile of the on-floor Concierge welcome the guest to the nineteenth floor.

> 耀眼的大理石、光亮的木质嵌板、华丽的帷幔和门口面带微笑的迎宾员会将您带到19楼。

> Convention floor plan

> 会议室平面图

1. Stage	1. 舞台
2. Concert hall	2. 音乐厅
3. Salon B	3. B沙龙
4. Foyer	4. 门厅
5. Kitchen	5. 厨房
6. Ballroom	6. 宴会厅
7. Salon A	7. A沙龙
8. Ontario Room	8. 安大略宴会厅
9. Toronto Room	9. 多伦多宴会厅
10. Canadian Room	10. 加拿大宴会厅

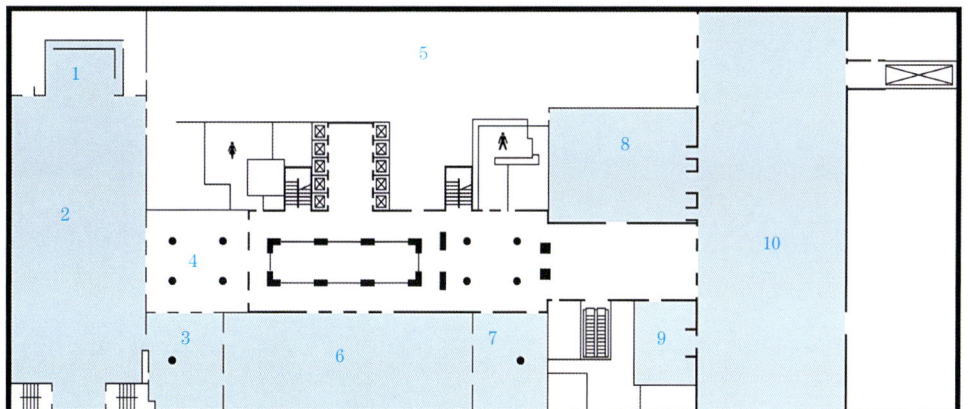

> The ballroom is a choice destination for corporate affairs and establishment weddings.
> The ballroom features its grand painting.
> Recently renovated, this elegant ballroom is more modern.

> 这间宴会厅是召开公司会议和举办订婚典礼的绝佳之选。

> 宴会厅中最醒目的要数挂在墙上的巨幅壁画。

> 最近刚刚翻新的宴会厅更具现代气息，十分优雅。

> The concert hall has a soaring ceiling, full stage and balcony, crystal chandelier, and arched windows.
> The concert hall can hold 1,085 for cocktails or 770 for dinner.
> The Ballroom has an Italianate palette, materials, and oil paintings inspired by mythological themes.

> 音乐厅拥有高高的屋顶、宽阔的舞台和阳台、水晶吊灯和拱形窗户。

> 音乐厅可举办1,085人的鸡尾酒会或770人的晚宴。

> 宴会厅采用意大利风格的色调、材料和神话主题的油画作装饰。

> The Québec Room has printing carpet and grand painting.
> The Ballroom has an eye-catching ceiling that boasts a fresco of clouds and chariots, topping off 25-foot windows and sprawling hardwood floors.
> The ballroom has been the set for many movies. It is one of the most photographed rooms in Toronto.

> 魁北克宴会厅采用印花地毯和巨幅壁画作装饰。
> 宴会厅有一个光彩夺目的壁画天花板，上面画着云朵和战车等图案；有高耸的窗户，窗户高25英尺；还有遍布整个房间的硬木地板。
> 这间宴会厅曾是众多电影的拍摄场地，也是多伦多最佳的取景地之一。

> The Salon on the nineteenth floor was designed in consultation with senior meeting planners and top level executives.
> The lounge has wooden wall and fireplace.

>19楼的宴会沙龙是设计师与高级会议策划师和酒店高管们共同协商后的杰作。
>休息大厅采用木质墙壁和壁炉作装饰。

> Deluxe Rooms are larger open-concept rooms, with a sitting area as well as a sleeping area.
> Located throughout the hotel, Fairmont Rooms are traditional guestrooms, with elegant, comfortable decor.
> With their larger sizes, Deluxe Rooms are ideal for families, or any guest who will be staying for an extended period.

> 豪华客房采用开放式设计，拥有一间客厅和一间卧室。
> 豪华客房极为宽敞，是家庭游客和渴望长期入住的游客的理想下榻之选。
> 遍布酒店各个楼层的费尔蒙特客房是传统的客房，装饰优雅、舒适。

THE FAIRMONT SAN FRANCISCO

费尔蒙特旧金山酒店

Completion date (项目建成时间): 2009 / Location (项目地点): San Francisco, USA / Designer (设计师): Champalimaud / Photographer (摄影师): Matthew Millman / Area (室内面积): 41,000 sqm

Centrally located, The Fairmont San Francisco is a short cable-car-trip from the bustling Downtown, Financial District, Union Square and Fisherman's Wharf.

The Fairmont San Francisco united a five-star design team led by Champalimaud to revitalise the iconic Penthouse as the world's ultimate suite. Created in the Roaring Twenties by famed American archeologist and art historian, Arthur Upham Pope, and spanning the entire eighth floor of the historic Main Building of The Fairmont San Francisco, the Penthouse offers 6,000 square feet of luxury. Arthur Upham Pope, a pioneering scholar on Persian art and architecture, designed The Penthouse and his exotic influence can be found in striking details throughout the suite.

The suite features three large bedrooms, a living room with grand piano, a formal dining room seating 60 people, a kitchen, a two-storey circular library crowned by a rotunda where a celestial map is rendered in gold leaf against a sapphire sky, a billiard room covered in Persian tile from floor to vaulted ceiling, and an expansive terrace with sweeping views of San Francisco. A secret passageway concealed behind bookshelves on the library's second floor lends a sense of intrigue to the fabled suite. The impressive art collections feature original works by David Hockney and other contemporary artists as well as a grouping of exquisite Chinese porcelain vases.

Champalimaud created a luxurious marriage of East meets West in The Penthouse, utilising the Moorish influence of the billiard room and terrace as its inspiration and adding dramatic design elements such as a hand-painted silver and black Chinoiserie wallcovering to the palatial dining room. Modern sophistication and a transitional mix of styles will combine with storied old-world glamour to create a suite of timeless elegance.

Recently restored to their original splendor, these Fairmont San Francisco hotel rooms are cosy and quiet. They are beautifully decorated with royal blues and golds, and each has a marble bath.

> Located in the hotel's Lobby, Laurel Court Restaurant & Bar has a high dome.
> The glamorous lobby has vaulted ceilings, Corinthian columns, a spectacular spiral staircase, and Rococo furniture.
> Dorothy Draper's design has been stripped away to reveal pristine marble floors and Corinthian columns trimmed in gold.

> 坐落于酒店大堂的Laurel Court餐厅与酒吧拥有高高的圆屋顶。
> 迷人的大厅拥有拱形的屋顶、科林斯石柱、壮观的旋转楼梯和洛可可式的家具。
> 原始的大理石地板和金色图纹镶嵌的科林斯石柱掩盖住了桃乐茜·德雷帕的设计。

费尔蒙特旧金山酒店坐落于旧金山市中心，从旧金山市中心的金融中心、联合广场和渔人码头坐缆车仅几分钟的路程。

费尔蒙特旧金山酒店联合由Champalimaud领衔的五星级设计团队，将酒店标志性的阁楼打造成了世界级的顶尖套房。这个阁楼由美国著名的考古学家和美术史学家——亚瑟·阿伯翰·蒲伯创作，占据了费尔蒙特旧金山酒店这栋历史建筑的整个八楼空间，6,000平方米的空间装饰奢华。波斯艺术与建筑的领先学者亚瑟·阿伯翰·蒲伯设计了这个顶层公寓，每间套房内令人叹为观止的设计中都贯穿着蒲伯的异国情愫。

套房中有3张大床、一间摆放着1架大钢琴的客厅、一间能容纳60人的正式餐厅、一个厨房、一个双层图书馆，图书馆上方是一个圆形大厅，大厅内，一片片金箔在蔚蓝的天空背景下拼成了一个天体图，此外，还有一个台球室和一个广阔的阳台，台球室内，从地面到屋顶全部采用波斯瓷砖铺设；阳台上，可一览旧金山的美丽景色。图书馆二层的书架后方隐藏着一个秘密通道，增加了传说中套房的神秘感。套房内的艺术品主要是大卫·霍克尼和其他一些现代艺术家的原创作品，还有一组高雅的中国瓷瓶。

Champalimaud的建筑师们在这个顶层公寓中打造了东西方相互融合的奢华组合，他们充分利用了台球室与阳台的摩尔文化氛围，并为宫殿般的餐厅增加了夸张的设计元素，如手工绘制的黑白色中国风壁纸等。现代的高超技术与各种风格的过渡性组合将与古典魅力一同打造一个凝聚着永恒优雅的套房。

这些费尔蒙特酒店的客房最近刚刚恢复其原始面貌，既温馨舒适又宁静典雅。酒店采用宝蓝色和金色为主色调，装修精美，每一间客房内都配备一个大理石浴盆。

> The wall coverings are a subdued, gold damask pattern, and the carpet has accents of royal and light blue, periwinkle, and buttercup yellow.
> Gold Room glitters with French provincial detailing. Ornate, gold-leaf bas reliefs decorate the walls.
> Light and airy, with an elaborately ornamented 22-foot ceiling, the Ballroom is stunning.

> 墙面涂料上镶嵌着压制的金色花锻图案，地毯上则印着淡蓝色的皇室长春花图案和毛茛黄图案。

> 采用法国地方风格装饰的Gold Room宴会厅熠熠生辉。宴会厅的墙壁采用装饰性的金叶浅浮雕作装饰。

> 这间宴会厅明亮通风，拥有22英尺高的屋顶，屋顶装饰精美，堪称美妙绝伦。

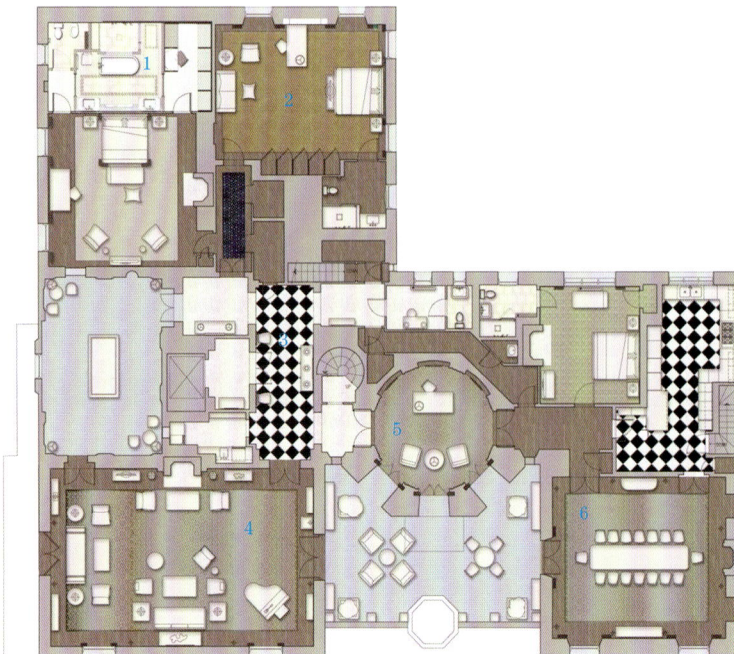

> Penthouse Suite plan
> 顶楼套房平面图

1. Bathroom	1. 浴室
2. Bedroom	2. 卧室
3. Corridor	3. 走廊
4. Reception room	4. 会客室
5. Study	5. 书房
6. Dining room	6. 餐厅

> The wraparound windows reveal a sweeping 270-degree panorama of the Golden Gate and Bay Bridges, Coit Tower, and Downtown.
> Crystal chandeliers, fireplace and mirror in the wall add authentic turn-of-the-century accents.
> The small meeting room is elegant with floral carpet.

> 弧形的窗户拥有270度的广阔视角，可一览金门大桥、科伊特塔和市中心的美妙景色。

> 水晶吊灯、壁炉和墙上的镜子增加了世纪之交时期的装饰韵味。

> 小型会议室的地面铺着花朵图案的地毯，看上去十分优雅。

> Main Building Suite has a spacious bedroom.
> Main Building Balcony Suite features an ornately filigreed terrace.
> The Fairmont Suite features its contemporary design.
> The bedroom of Penthouse Suite has its characteristic with the map wall.

> 主建筑的套房内拥有一间宽敞的卧室。
> 主建筑的阳台套房拥有一个采用金银细线工装饰的阳台。
> 费尔蒙特套房以现代风格的设计为特色。
> 顶楼套房的卧室以地图墙为特色。

Index 索引

19. The St.Regis Washington, D.C
    Washington, D.C, USA
    ranu@mcc-pr.com
    Sills Huniford

20. Carlton Baglioni Hotel
    Milano, Italy
    press@baglionihotels.com
    Reggiori

21. Regina Hotel Baglioni
    Rome, Italy
    press@baglionihotels.com
    Carlo Busiri Vici

22. The Landmark London
    London, UK
    webpr@thelandmark.co.uk
    HBA

23. Hotel Excelsior
    Munich, Germany
    reception@hotelexcelsior-frankfurt.de
    Jochen Dahms, Atelier Dahms, Tauberbischofsheim

24. Hotel Koenigshof
    München, Germany
    res.koenigshof@geisel-privathotels.de
    Jochen Dahms, Atelier Dahms, Tauberbischofsheim

25. Grand Hotel Majestic Già Baglioni
    Bologna, Italy
    salesghmajestic@duetorrihotels.com
    Alfonso Torreggiani

26. Hotel Cipriani
    Venice, Italy
    joshuashi@126.com
    Michel Jouannet

27. Grand Hotel Excelsior Vittoria
    Sorrento, Italy
    exvitt@exvitt.it

    Richard Kerr, Colin Gold
28. Relais Santa Croce
    Florence, Italy
    press@baglionihotels.com
    Vasari

29. San Clemente Palace Hotel
    Venice, Italy
    marketing@sanclementepalacevenice.com
    Carlo Busiri Vici

30. The Fairmont Copley Plaza Hotel
    Boston, USA
    boston@fairmont.com
    Jinnie Kim Design

31. The Fairmont Hotel Macdonald
    Edmonton, Canada
    hotelmacdonald@fairmont.com
    Heather Jones & Associates

32. Fairmont Royal York
    Toronto, Canada
    royalyorkhotel@fairmont.com
    Forrest Perkins

33. The Fairmont San Francisco
    San Francisco, USA
    innas@champalimauddesign.com
    Champalimaud

图书在版编目（CIP）数据

新古典风格酒店 / 马竹音编；代伟楠译. ——沈阳
：辽宁科学技术出版社，2012.4
　ISBN 978-7-5381-7337-6

　I. ①新… II. ①马…　②代… III. ①饭店—建筑设
计—欧洲—图集 IV. ①TU247.4-64

　中国版本图书馆CIP数据核字（2012）第011256号

出版发行：辽宁科学技术出版社
　　　　　（地址：沈阳市和平区十一纬路29号　邮编：110003）
印 刷 者：利丰雅高印刷（深圳）有限公司
经 销 者：各地新华书店
幅面尺寸：230mm×290mm
印　　张：48
插　　页：4
字　　数：60千字
印　　数：1~2000
出版时间：2012年4月第1版
印刷时间：2012年4月第1次印刷
责任编辑：陈慈良
封面设计：曹　琳
版式设计：曹　琳
责任校对：周　文

书　　号：ISBN 978-7-5381-7337-6
定　　价：328.00元

联系电话：024-23284360
邮购热线：024-23284502
E-mail: lnkjc@126.com
http://www.lnkj.com.cn
本书网址：www.lnkj.cn/uri.sh/7337